WHEN THE CLOCK STRUCK ZERO

PROFESSOR JOHN TAYLOR

WHEN THE CLOCK STRUCK ZERO

Science's Ultimate Limits

St. Martin's Press
New York

The illustrations which appear on pages 131 and 165 of this book are taken from *Brain, Mind, and Behaviour*, Second Edition, by Floyd E. Bloom and Arlyne Lazerson, copyright © 1988 by Educational Broadcasting Corporation, and are reprinted by permission of W. H. Freeman and Company.

Library of Congress Cataloging-in-Publication Data

Taylor, John Gerald
When the clock struck zero / J.G. Taylor.
p. cm.
"A Thomas Dunne book."
ISBN 0-312-11065-0
1. Cosmology—Popular works. 2. Evolution—Popular works. I. Title.
QB982.T39 1994
523.1—dc20 94-7142
CIP

First published in Great Britain by Picador.

First U.S. Edition: August 1994
10 9 8 7 6 5 4 3 2 1

CONTENTS

CONTENTS

PREFACE

I HAVE LIVED my whole working life as a scientist. As a boy I lived in a household where science was regarded as an important way of looking at the world. As I grew up I realized ever more strongly that science was the only way to probe the mystery of the Universe and obtain definite answers to questions on the nature of existence. In order to do that myself I have tried to understand and investigate a number of fields – particle physics, cosmology (especially black holes), nuclear physics, solid-state physics, neural modelling and brain sciences, artificial neural networks and the non-linear dynamics of complex systems.

More recently I wished to take stock of the certainties which science could provide in an uncertain world. What was the bedrock of existence which the scientific mode of thought could be expected to uncover? To discover this, about two years ago I went on an odyssey to discuss with distinguished scientific colleagues what such a bedrock might be. This resulted in a sharpened sense of my own ideas and understandings, leading to this book, as a culmination of nearly forty working years in scientific research.

I have narrowed down the search to the attempt to answer four basic questions. These questions – How did the Universe begin? Why is there something rather than nothing? Why *that* theory of everything? What about consciousness? – are regarded by some as impossible to answer. Yet we must see how far we can get in tackling them if we want to make sense of life using as much of science as we can. So it is to answer these questions in the spirit of science that this book is really addressed. Rather like the fine print in an insurance or other legal document these problems point out that the theory of everything, claimed by some to be just around the corner, may actually be rather limited. It may only apply to a special class of situation. This may not be relevant when we wish to understand the Universe so as to see our place in it and the total order existing. Thus we have to try to answer some impossible questions that seem beyond the realm of the standard theory of everything. This is indicated by the title of the book. For how can a clock strike zero? Clocks strike one or twelve or, exceptionally, thirteen. How can one hear zero chimes of a clock? Who could be subtle enough to create such a timepiece? And would there have been any time at time zero?

This book is concerned with analysing whether science is really up to such things or, if not, what the limitations seem to be. In the process we will see that the claims of 'theory-of-everything' fanatics like Stephen Hawking are most probably wrong (once again – he had previously predicted one such theory of everything in 1979 which was proved wrong three years later). We shall also see that the supposed 'Mind of God' being glimpsed in recent scientific advances is an illusion. For 'Mind' may not be such a mysterious phenomenon after all. Science will be able to give an explanation of the mental world, very likely along the lines suggested later in the book, associated with a general 'relational theory of mind'. Furthermore, scientific evidence even points to there being no ultimate theory of everything, and perhaps no beginning to the Universe. I will make the limitations of science clearer as we progress through the book, in particular that science will never be able to give significance to life. The implications for ourselves – especially the challenge this presents to us in discovering how to find significance – will be spelt out in the conclusion.

I can only claim for myself the responsibility of my main theses and how these are developed. Any errors or mistakes are mine alone. But I must thank the myriad of my distinguished scientific colleagues who have had so much patience with my stumbling attempts to understand the Universe in its rich panoply. I must also thank my wife, Pam, and my daughter, Elizabeth, for their excellent and trenchant comments on early versions of the book. Although they disagree with my main tenets, without their help so many of my thoughts would have remained incomprehensible when expressed on paper. I must also thank my wife for her excellent typing of the book.

WHEN THE CLOCK STRUCK ZERO

PART ONE INTRODUCTION

Chapter One

WHAT DOES IT ALL MEAN?

'I'm afraid of the Universe – it's a lot bigger than me'

AT ONE TIME or another each of us has wanted to know how to understand the Universe and our place in it. When intimations of mortality arise on the death of someone in the family, or of a friend, we are driven to ask the questions 'Who am I?', 'Is there life after death?', 'What does it all mean?' and so on. At another time we might be outside in a terrifying storm and feel almost powerless to brave the fury of the elements as the wind shrieks about us, trying to dislodge us from the face of the earth. Or we look up at night and see a carpet of stars glimmering mysteriously across the sky. We then recognize the vastness of space, the power of the elements, and the smallness and fragility of our own personal world. On the other hand, each of us has our own internal thoughts and sensations. We seem to have complete freedom in this internal world, and we ascribe a similar mental existence to those we see around us. Our minds apparently exist somewhat independently of our bodies; we can thrill at the storm beating down on us in spite of being nearly swept away by it.

Our enormous range of experiences of the Universe has led to many different answers to the question as to how we can comprehend it. Some say that the existence of our inner mental world, or of the need for a creator of the whole complex, awe-inspiring Universe, gives clear evidence for the existence of a cosmic intelligence – God. Through the inner mental world personal contact with God may be made by prayer. This direct relation with a divine all-powerful entity can give great support to believers in times of tribulation. Communal worship can strengthen such beliefs, as can family and social indoctrination. The resulting belief system can be very supportive. Islamic fundamentalism, for example, can be seen as so effective because of the certainty it gives to its believers that theirs is the correct and only way of life, in comparison to the uncertainties of unbelievers.

There are also those who attempt to live their lives without making such a leap of faith. Their attitudes vary from those of agnostics, who would try only to use reason to face up to the nature of life and the Universe but who are open as to the existence or otherwise of God, to those of atheists who deny that God exists. A good example of the reasoning of these latter is that of the biologist the late Sir Peter Medawar, who wrote: 'The price in blood and tears that mankind generally has had to pay for the comfort and spiritual refreshment that religion has brought a few has been too great to justify our entrusting moral accountancy to religious belief.'

Similar but more cosmic comments have been made by the cosmologist Stephen Hawking: 'We are insignificant creatures on a minor planet of a very average star in the outer suburbs of one of a hundred thousand million galaxies. So it is difficult to believe in a God that would care about us or even notice our existence.' When he later used the oft-quoted phrase 'the Mind of God', it was unlikely that he was really turning back to religion.

All such approaches could be said to be rationalist, in that it is hoped to understand ourselves and the Universe only by reason and not by acts of faith.

There are also people still searching for meaning in life but who have not yet found it. They may just have drifted along as creatures of habit, accepting the customs of their forebears without much thought or action. They are in the 'don't yet know' category when it comes to religious or rationalist guidance.

Both the rationalist and the undecided person are still none the wiser as to the greater purpose of things. Some, such as Medawar quoted above, might even claim that there is no way scientifically ever to know of this greater purpose; as he says: '. . . it is logically outside the competence of science to answer questions to do with first and last things.'

Yet that attitude can be hard to live with. There is still a gap in many lives due to this lack of meaning. Carl Jung, the psychologist, wrote that 'among all my patients in the second half of life (over 35) there has not been one whose problem in the last resort was not that of finding a religious outlook on life.' Seen from a cosmical level we on earth may well be puny and insignificant. But we are certainly not so unimportant when observed from our own level. From inside ourselves we are the whole world. How can we reconcile these apparently contradictory existences: a warm internal one filled with our own needs, emotions and desires, versus our alien external materialistic world. We cry out for guidance! This inability of science to give any sense of purpose to life has led to a recent set of attacks on the scientific approach in toto. As recently claimed by Bryan Appleyard, the journalist son of a physicist, 'science is not a neutral or innocent commodity which can be employed as a convenience by people only wishing to partake of the West's material power . . . It is spiritually corrosive, burning away ancient authorities and traditions. It cannot really co-exist with anything. . . . As it burns away all competition, the question becomes: what kind of life is it that science offers to its people? . . . What does it tell us about ourselves and how we must live?'

No ready answer seems to be obtainable to that question. We appear to be in one or other of two extreme situations: either we accept science, but must acknowledge that it is limited and that there never can be a complete answer to help us comprehend the 'why' of the Universe and our place in it, or we make a leap of faith and believe one of the world's many religions. In the first, it is claimed that we must learn to live in what is basically a mechanical world. Our warm internal mental world is just an epiphenomenon arising from the complexity of activity in our brains. No light is cast on the emotions – love, hate, fear, anger. The world becomes a cold, inhuman place. Moreover, science can never answer 'why' questions, only 'how' ones. No reason can ever be discovered for existence; we will have to muddle through as best we can. In the second, the path of faith, all will be 'revealed', although if we press for detailed answers to these deep 'how' questions they tend not to be forthcoming. It is as if a veil has to be drawn across the altar of understanding promised by any particular religion if too much clarity is demanded.

Either way ahead has its drawbacks. Both appear to lead to curtailment of enquiry. I think we all must accept that religions draw a veil over the deepest understandings of life. But science also appears to have its limitations. In spite of proving ever more effective in making life around us comfortable it would seem not to be able to give answers to moral or aesthetic questions. That is the thrust of Appleyard's concern described above. Nor can it, from Medawar's quote earlier, answer first or last questions about the nature of the Universe. We seem, then, to be completely blocked in pulling back the veil covering the altar of understanding. How can we make progress in the face of these depressing conclusions? Radical recent developments seem to indicate that the situation may not be as bad as I have just described it.

Theories of Everything

RECENTLY THERE have been claims by some scientists that the end of fundamental science might be in sight and that a 'Theory of Everything', or TOE, is just around the corner. Stephen Hawking, for example, in a public lecture he gave in Jerusalem in 1988, said, 'It now looks quite hopeful that by the end of the century we may have a complete theory to describe the universe.' Other scientists have been making similar claims that such a theory can be partially glimpsed. The enormous strides taken in recent years in trying to unify (combine and simplify) the forces of nature (which are those of electromagnetism, radioactivity, gravity, and the strong nuclear force holding the neutrons and protons together inside the nucleus) give some support to this enormous optimism. The 'holy grail' of the ultimate all-unified theory of the forces of nature should, according to its protagonists, be such a final theory. There could be nothing finer, better, bigger or more beautiful in the manner of theories of nature.

A theory of everything should, so it is claimed, contain within itself the answer to all possible questions that one could ever ask about the Universe. It should even be able to answer all the impossible questions ever asked. Such deep questions as to how the Universe began, how it will end, the nature of Mind, and so on, are all expected to be answered by means of this 'theory to end all theories'. It has even been claimed that once it is discovered, God will no longer be needed. Even if the details of the answers to all these questions are not rapidly available their principal features (such as the nature of the unified force and the truly elementary particles) are expected to be. Science, after all, would appear to hold the key; our earlier comments about its limitations are claimed by the scientific optimists to be an illusion. The true reason for human existence would apparently become clear. We would all become atheists, and religious institutions would fade away. Monasteries and churches would dissolve themselves. The only fundamentalism that would be left would in the end be a scientific one. Aldous Huxley's 'Brave New World' would be many steps closer. The prevalent idea of the existence of free will would disappear. All criminal activity would be considered as caused by brain malfunction, so that criminals would no longer be locked up (only to be later released to continue raping, mugging or murdering) but would be hospitalized and treated as sick. Behaviour therapy or brain microsurgery sessions would replace prison sentences. Hedonism would flourish. It is claimed that many other changes would occur in a society resulting from such scientific knowledge.

But is the Universe really as simple as that? The rationalists in society would like to think so, especially the scientists espousing the new wave of scientific fundamentalism. They desire the Universe to be reducible in totality to a single theory. They would like to stare at the mathematical equation updating Einstein's famous '$E = MC^2$' to 'TOE = . . .' (to be inserted). The unravelling of the implications of the equation 'TOE' might take centuries, as is now becoming apparent for the latest TOE candidate, Superstrings, for which no physical predictions have yet been made. Yet the TOE would provide the security of certainty and absolute truth so far denied by science and unobtainable to those who are unprepared to take the leap of faith into a particular religion. But as Milton Beurle the American comedian said: 'I'm afraid of the Universe – it's a lot bigger than me.' Doesn't the claim for a unique TOE go beyond the bounds of reasonableness?

It may not, of course, be misplaced optimism that is driving scientists on to discover the TOE. In spite of the rashness of the claims being made about the TOE, it may indeed be true that after only three to four hundred years of travelling on it the end of the scientific road has been reached. This in spite of the fact that life has been on earth for about three billion years, and is expected to continue so (barring unnecessary accidents) for a similar length of time. What will humanity have to think about for such a long time ahead if we really have discovered the secret of existence? Will fundamental science disappear, rather like the teaching of Latin in schools, since it would no longer be deemed useful?

Recent remarks by some of the scientists and science journalists following the TOE bandwagon indicate, however, that they seem to have lost their scientific nerve. They are arguing that the elegant mathematical nature of the yet vaguely glimpsed TOE is a strong indication of an intelligence at work behind it all. The TOE is talked about as part of 'the Mind of God'. It is as if this group is after all searching for God and, although still wary of taking a religious-style leap of faith, are going to do it in spite of themselves. If there is such wavering between the two extremes of science and religion how can the non-expert in either make sense of it? The spectre is also raised of discovering both the TOE and God 'behind' it, as a considerable number of people today believe.

These questions have been agonized over, with thousands of books written containing answers claiming to be correct, over the last two thousand or more years. The claim to have a theory of everything was even made by the ancient Greek philosophers over two thousand five hundred years ago. For example, Thales claimed that all was made of water. So you would only need 'H_2O' on your slogan-emblazoned T-shirt if you were a Thalian.

One crucial new feature has entered the situation: the amazing progress in science over this century. Our whole society runs ever more smoothly on devices produced by scientific advances, even down to the word-processing techniques used in typing these very words and the computerized typesetting methods employed in printing the whole book. These developments are powered by a fantastically successful advance in uncovering the secrets of nature. Beautiful symmetries have been discerned in the most elementary of particles, and powerful and elegant mathematical theories have been constructed which correlate and develop these features.

The physical world now seems to have been built almost totally according to powerful mathematical rules. Moreover, the mathematical structure becoming apparent is being taken as an indication of the existence of an underlying theory of all-embracing comprehensiveness and strength.

Such an advance has been made possible only through strict adherence to the scientific method. Science looks at the facts, makes a conjecture or theory about them, deduces predictions from the theory and tests them until the theory seems well corroborated or has failed. If the latter possibility, then one must start with a new, hopefully better, conjectured theory. Testing a theory to destruction (falsifiability) involves discussion between scientists. It leads to objectively validated theories, and implies that there must be a common reality 'out there' which is being uncovered by the collective endeavours of the scientific community.

The scientific corpus of ideas is claimed to be different from all the other ideas about reality which mankind has developed up to now. Thales' idea mentioned above that 'All is made of water' was contested by others at the time and other proposals made – 'earth, air, fire and water', for example, was another candidate. The underlying nature of reality has been claimed by some philosophers to be purely mental, by others purely physical, and yet again a mixture of the two by a further group. The objective advance of science, tested carefully against the reality it is trying to explain, has led us forward, in the last few hundred years, to a physical world governed by some underlying beautifully elegant mathematical equations expressing the TOE. Real progress now appears to be occurring; unique answers to some of the deep questions raised earlier are, it is claimed, now being obtained. All humanity's curiosity will be satisfied in the end, it is being argued, when the TOE has been thoroughly unravelled and all its implications discovered. It will then seem that our earlier cynicism about the TOE was quite wrong.

Impossible questions?

ARE WE NOT in danger of being carried away by the new scientific optimists? There are further, even deeper questions which we will have to face up to before we should be satisfied. 'Everything' is a very large concept. Does a theory of 'everything' not have to describe where all objects are at any one time? And why *that* theory rather than any other? Does it not also have to answer how the Universe is here at all? In other words, why is there something rather than nothing? There is that part of reality associated with each of our inner mental worlds, that of experiences, sensations, or 'raw feels'. How does that arise? Or is consciousness to be regarded as a primitive concept, something which cannot be probed further, as some philosophers would claim? Any theory of everything which leaves out this realm of experience would not be worthy of the name. Not much attention appears, as yet, to have been paid to the Mind by the TOE enthusiasts.

In summary, then, we have to see how far science can go in answering, among others, the fundamental questions:

How did the Universe begin?
Why is there something rather than nothing?
How can we explain all the complexity in the world about us?
Why that particular theory of everything?
How can we explain consciousness?

Chapter Two

QUESTIONS OF SCIENCE

'Never say can't.'

IT MAY HAVE been an early craving for superhuman capabilities, experienced by many as they grow to adulthood, which has led to belief in the supernatural. Magic and witchcraft appear potential ways of gaining the upper hand over nature and one's fellow beings. Supernatural forces are thought to exist which are capable of being tapped by special techniques or by especially disturbed people, as in poltergeist cases.

While the claims made have not stood up to attempts at detailed scientific investigation, some still hope they are valid. Notions of witchcraft, magic, astrology and other supernatural demonic-possession phenomena still permeate society at all levels. For example, the first part of a newspaper read by a good proportion of people today is the astrology section. Faith healing and mediumship still flourish in society. The lack of validation of supernatural phenomena has not, in any case, affected belief in a more general feature of a supernatural kind, that of the existence of God.

Credence in the existence of a single or multiple deity who is at one time many things – all-powerful, all-good, creator of heaven and earth, sustainer of the Universe – is an almost universal phenomenon in human societies and civilizations. God may be seen as a personification of those powers which we would wish for ourselves. He is prayed to so as to use his powers for our good, or to direct other harmful forces away from us. 'Give us this day our daily bread' and '. . . deliver us from evil' are said by many throughout the world every Sunday. The Islamic jihad or holy war is another modern instance of having God on one's own side.

Belief in God would also seem to involve the believer in keeping to what my mother often demanded of me when I was young: 'Never say can't.' She meant this as an encouragement to me when I was facing up to an apparently impossible task. Nothing was impossible to her. That would seem to be the same in the case of religious believers, for how otherwise can they reconcile apparently contradictory features of the God being believed in? Thus he is all-powerful, but yet cannot seem to prevent misfortune befalling those who live apparently blameless lives. How could God allow an infant to die from leukaemia or AIDS if he is all-good and all-powerful? How can he allow the dictator of Iraq to commit genocide of his own people with apparent impunity? How could God have allowed Hitler's slaughter of the Jews? Such pain and suffering as abound in the world do sometimes cause people to lose their faith, yet not always. In spite of these terrible contradictions religious believers still maintain their faith and thus the propensity never to say 'can't'.

Is that, then, all that is needed to surmount all obstacles – essentially by bringing in a set of principles which may conflict and then blithely ignoring the contradictions? But that misses the original point. For ignoring inconsistencies does not resolve the problems they pose. They are still there to be faced up to. Moreover this evasion can induce a tendency to concentrate on other quite difficult features, so as to further avoid noticing the contradictions. There are indeed hard problems ahead in understanding the material world, which appear to lead to questions that seem almost, if not completely, impossible. But we must be careful to recognize them as different from the original impossible questions (the nature of Mind, creation of the Universe, etc.) raised in Chapter 1. We must not be sidetracked by pursuing these new questions. Puzzling over them, especially as some are decidedly non-trivial, takes time and human effort. But we would be in danger of being distracted from the much deeper problems of the nature of life, the Universe and some crucial other ideas.

Let us look at some of these difficult problems masquerading as ultimate questions before we move on.

Difficult problems

THERE ARE MANY difficult problems facing science that are distinct from those which arise in attempting to discover and test any theory of everything. The so-called 'cutting edge' of science is usually thought of as involving investigations of the very large – cosmology, astrophysics – or the very small – high-energy or elementary-particle physics. Yet over the vast range of distances in between – from the size of a nucleus up to the size of our galaxy and beyond – there are enormously hard problems which only now are we beginning to appreciate as intellectually challenging. These involve the complex systems we are surrounded by in our daily lives, for example: solid bodies, some with special (superconducting) properties, fast-flowing liquids, and moving clouds of water vapour usually called weather. I will focus here on just these three types of systems to indicate the richness of the problems they present to science. At the same time I want to use them as examples of typical scientific problems at the turn of the millennium which throw light on the nature of science and its limitations.

Superconductivity, as its name implies, is the ability of some materials to allow a kind of perpetual motion (conduction) of electric current in their interior. It usually occurs in metals or alloys (mixtures of two or more metals to increase their strength) when they are cooled down far enough. Thus below minus 266 degrees Centigrade the electrical resistance of lead completely disappears, and an electric current can be kept flowing unattenuated in a ring of the metal (and has been for several years in some cases). Due to its vanishing resistance, very high electrical currents can flow through such material. It can therefore be used to generate extremely powerful magnetic fields when used in the windings of an electromagnet. Such superconducting magnets could be used in the near future in magnetically levitated trains, running in a perfectly frictionless manner. Highly efficient transmission of electric power along superconducting power lines is also an important application being predicted for this new device.

The use of such superconducting materials is hampered by the need to cool them to very low temperatures, which is both costly and requires very specialized equipment. The search for material superconducting at higher temperatures, if possible even at room temperature, was sparked off in 1986 by the discovery by George Bednorz and Alex Meller in Zurich (for which they received a Nobel prize within a year) of a completely new type of material which was superconducting at about 30 degrees higher than lead. Since then there have been more than twenty thousand scientific publications with reports of new superconducting substances and ever higher temperatures; the record is at about 120 degrees above that for lead and could lead to the breakthrough in the applications mentioned earlier.

In spite of this welter of experimental reports, and substances exhibiting high-temperature superconductivity, no firmly established theory has yet emerged (which makes the Bednorz and Meller discovery of the new material even more remarkable as a 'shot in the dark' which worked). The challenge is that, in spite of knowing the underlying laws governing the atoms of any material, the way these laws determine the high-temperature superconducting properties of various sorts of new materials is hard to ascertain because of the complexity of the particular sets of atoms in question. But though a hard problem, it does not seem an impossible one, especially since the superconductivity of lead has been understood for about thirty years.

A similar problem arises in understanding the phenomenon of turbulence in the flow of fluids. This arises when a fluid flows along a tube at too high a speed for the motion to be along straight lines. The fluid breaks up and appears to have a high degree of complexity and unpredictability but still possesses a certain degree of coherence. The basic equations of motion of the fluid are known. Computer simulations of fast-flowing liquids, starting in the early 1980s, showed the presence of elongated filaments of liquid embedded in an immense disorder. These filaments were observed later in an experiment on fluid flow brought about by counter-rotating disks at opposite ends of a cylinder. This has led to a quite successful 'two-fluid model' of turbulent flow, in which one fluid is composed of the elongated filament and the other of the disordered background in which it floats.

A further hard problem in science is that of weather forecasting. This problem is exacerbated by the type of randomness seen in turbulence in that one's success can be tested by going out without an umbrella on the basis of a good forecast and coming home soaked to the skin from an unexpected downpour. The trail of destruction left by the unpredicted storm across France and the UK in October 1987 is difficult for the weather men to live down. The UK Meteorological Office forecast a similar vortex of mayhem over Somerset on Monday 29 October 1991, but were left most embarrassed by a bright, clear and calm Monday and derisive pictures in the following morning's newspapers of lightly clad models strolling along sunlit beaches. The forecasters are as scientific as they can be, using a newly installed computer called a Cray YMR 8132 running at 2000 million calculations per second. In the older, slower machine 50,000 weather observations a day (gathered from all over the world) were used to produce, in four minutes, a global forecast of the weather for fifteen minutes ahead and longer-term predictions; the Cray works similarly.

At present it is claimed that in the past five years real accuracy has improved from predictions of the weather two days hence to three days due to a better model of the climate, bigger and faster computers and improved weather observations; the Cray adds another day of prediction. But there may be diminishing returns; it may cost ten times more to compute an extra day's weather and one hundred times more for two further days. However, even if there were no cash limit the 'butterfly effect', in which the flapping of a butterfly's wings in Tokyo could cause a storm in New York, would stop much further progress. This arises due to the extreme sensitivity, in the equations being used to predict the weather, of the conditions at one time on the observations at an earlier one. This brings in the fashionable subject of 'chaos', first noticed when an American meteorologist discovered that a very small change in an initial state of affairs could produce wild, unpredictable fluctuations later.

The occurrence of chaos puts a limit on the degree of predictability possible from purely deterministic equations. It used to be thought that the entire history of the Universe, past and future, is determined once you know, at one instance of time, exactly where each particle is and how fast it is moving. This determinism is clearly expressed in the writings of the French eighteenth-century mathematician Laplace:

> An intellect which at any given moment knew all the forces that animate Nature and the mutual positions of the beings that comprise it, if this intellect were vast enough to submit its data to analysis, could condense into a single formula the movement of the greatest bodies of the Universe and that of the lightest atom: for such an intellect nothing could be uncertain; and the future just like the past would be present before its eyes.

Laplace's fatalist claim is now in serious trouble. Chaos indicates that, at least for some non-linear systems including that of the weather, accurate predictions too far into the future would be impossible. For example, it has been suggested that due to these chaotic effects an accurate weather forecast for only about ten days ahead could be possible. Even if the laws of nature were deterministic the enormous sensitivity to initial conditions (and to any background noise in the intellect's computer) would preclude predictability at the level claimed by Laplace.

It has recently been proposed by some chaos practitioners that science has been systematically misleading us for the last three hundred years; we now see that the future may be fixed but we could never compute it. It has even been suggested that free will may have its origin in this unpredictability. For the present let us note that the impossibility of making predictions is a limit which does not affect the underlying laws of physics, but only our ability to understand what they imply in some complex situations.

Chaos is important for the science of complex systems and is leading to a better understanding of these systems. But it does not change the underlying laws such as Newton's laws of classical motion, or those of quantum mechanics, or Einstein's theories of relativity. Chaos does mean that in going between levels – from simple structures to complex – completely new and unexpected phenomena may arise. It does not, however, mean that the more complex level cannot be deduced from the simpler, but only that it is far more complicated than had been previously suspected. Chaotic patterns discovered in the weather and in turbulence are being analysed mathematically and chaos can be predicted for given systems. That is not a new situation in science. Beautiful patterns of self-organization were noticed in chemical reactions, but only later understood in terms of detailed dynamical models for the underlying constituents. A new dimension has been added to scientific understanding by chaos, but not a completely new direction. There is no barrier between different levels – the Universe may still be seen in a grain of sand, as the poet William Blake eloquently suggested:

> To see a World in a Grain of Sand,
> And a Heaven in a Wild Flower,
> Hold Infinity in the palm of your hand,
> And Eternity in an hour.

Impossible questions

I DESCRIBED IN Chapter 1 the set of apparently impossible questions which I feel are crucial in coming to terms with the Universe. Taking account of our discussion of the nature of complexity and chaos in the previous section, the basic set of problems I wish to consider in this book are:

How was the Universe created?
Why is that TOE the particular one?

How can we explain consciousness?
Why is there something rather than nothing?

There are other questions which might be added to this list. 'How will the Universe end?' and 'How did time begin?' are the next two on the list associated with the material world, while 'How did life begin?' seems as difficult as our question on consciousness for mental phenomena. Let us suppose that answers to the above 'impossible problems' are obtained, or further natural developments in present scientific understanding occur. Then none of these further questions is impossible to answer. Let me justify that claim. If we have understood how the Universe is evolving now then we should be able, in principle, to extend the application of the scientific laws used in that answer to follow the later evolution of the Universe. If this development can be followed indefinitely we should be led to a scenario for the end of the Universe. There could be enormous practical limitations to this due to the possibly chaotic nature inherent in the model that has been developed embodying the evolution of the Universe; the 'butterfly effect' discussed earlier, in which the flapping of a butterfly's wings in Tokyo could change the weather in New York, is at the basis of this. Small effects in an early stage of the history of the Universe could be magnified so much that they would make a prediction of the precise state of the Universe later increasingly hazardous. In spite of that, some general or averaged features of the later development could be expected to be foreseen, such as the continued expansion of the Universe, with the galaxies still streaming away from each other (assuming all Matter is at the same temperature), and so on.

I should emphasize that our ability to foresee the fate of the Universe may only depend on our present comprehension of it. It would not seem in general to be a prerequisite that we simultaneously solve the question of the Creation of the Universe. It would undoubtedly improve the credence we give to ideas we have as to how the Universe is evolving if these ideas were able, at the same time, to explain other more difficult phenomena, such as the beginning of the Universe. If that is possible it would be a benefit, but it is really a question of levels of explanation. Any level of explanation of the ultimate fate of the Universe is justified provided that the ideas and models used in foreseeing the future, based on that level, only lead to conditions – say of pressure and temperature – in which we know or expect them to be valid. Otherwise other levels of explanation will have to be employed.

Suppose, for instance, it was predicted from our model of the present Universe that there would be increasing compression and temperatures would increase without limit. We would then be required to go to an ever-deeper level of modelling involving higher temperatures to take account of the increasingly fiery end. Such a problem would arise if an initial expanding phase of the Universe were to develop into a contracting one. At some point in the future the average temperature of the Universe would become so high that the laws of classical physics would have to be replaced by those of quantum physics – a clear change of level of explanation. Later still, those laws themselves would become modified. One could also expect, in terms of the symmetry between the expanding beginning and the fiery end, that an understanding of the ultimate fate of the Universe could require the same levels of explanation as does its creation. We should expect that understanding the beginning of the Universe taps all possible levels of explanation. Thus the end would be in the beginning. On the other hand, if the Universe expanded for ever, then its present behaviour might be expected to be extended to describe ever later times.

Other questions about the material world, such as 'How did time begin?' which was raised earlier in this chapter, are ones we would expect to be answered as part of the main question on the beginning of the Universe. Time, space and Matter are expected to have been created at that event. To call it the 'initial' event would be to assume that there was a cosmic clock already ticking away. Both time and space are themselves to be regarded as dynamical objects in modern relativity theory. In any case we cannot leave a cosmic clock ticking away in the background before the Universe emerged, for we would be left with the question 'Who created the clock?'

Developments over the last few decades indicate that the creation of life is not an impossible problem. The rules for the construction of self-replicating systems have begun to be formulated and insights have arisen from computer simulations of such systems. The key question has been reduced from studies of proteins and DNA to the creation of a self-reproducing catalyst for it (called RNA). This latter would be a molecule with the ability both to reproduce itself and to make further RNA molecules which can catalyse the building of further molecules, and so on. An RNA molecule has already been created which can make small RNA molecules one-tenth of its size. Genetic evolution is also a complicated feature of the problem, but one which should not be impossible to resolve in terms of present science.

We have now shown that some of the four impossible questions are crucial ones in order to answer other important 'impossible' questions, and that further questions are hard but not impossible. Let us now consider briefly the nature of these four 'impossible' questions themselves. In particular we will discuss in what manner, if any, they are outside science.

In 1928 the American astronomer Edwin Hubble published results based on measurements of the Doppler shifts of distant galaxies, that the galaxies were all streaming away from each other. If their motion were to be extrapolated backwards it would imply that about fifteen thousand million years ago they had all been highly compressed, and had expanded at an increasing speed since then. In 1964 relic radiation of the early, hot stage was discovered to be bathing the Universe. These and related results enabled understanding of the early phase of the Universe to be pushed back to only one-hundredth of a second or so from its beginnings. What about even earlier? That seems still a scientific question, although made difficult by financial limits to probe matter at ever higher temperatures and compression, either here on earth in ever higher-energy particle accelerators, or in the heavens seen through larger and more efficient telescopes.

Even tentative answers to the question as to how the Universe began have been proposed. They are based on a very important ingredient of the new quantum or wave mechanics which replaced Newton's three famous laws of motion. This crucial feature is that events may only happen with a probability and are not a certainty. As such, there is a non-zero probability of, say, a particle such as an electron appearing out of the vacuum. In fact a vacuum is full of possibilities, one of which is the appearance of the Universe itself. It had been created from nothing, as it were. As one American scientist graphically described it, 'The Universe is a free lunch.' The details of how this free lunch might have come about will be considered later, but it is already clear that the ideas can, in principle, lead to experimentally testable predictions. It therefore appears to be within the bounds of science.

The next question is why that particular theory was required to be used in describing the creation of the Universe from nothing. This is a reformulation of 'Why that TOE?' incorporating the latest and best scientific knowledge. On the surface, that question would appear to have no scientific way of implementation. A TOE would, by its very nature, be able to answer all scientific questions about the nature of the Universe. But it would appear not to be able to explain itself. In response to such a viewpoint the scientific optimists would claim, with some justification, that all scientific questions – those leading to experimentally testable predictions – should be answerable by the TOE. In that scientific way one could probe the TOE's justifications. It could therefore be argued that scientific grounds could be obtained to answer the question 'Why that TOE?' It is because it fits all the scientifically known facts. More will be said about this later, but at least the question appears to be inside the scientific arena.

One of the facts of existence is that we are conscious and carry our own mental world about with us. A feature left out of detailed accounts by the scientific optimists and other proponents of the TOE is that it seems hard to give a scientific answer, based on any TOE presently known, to explain the detailed nature of this consciousness. In spite of the enormous efforts of psychologists and neuro-researchers, some would claim that this question will always lie outside the scientific pale. For example, the distinguished British psychologist Richard Gregory said in a recent radio programme that the inner nature of consciousness – the explanation of the so-called 'raw feels' of the experiences in our minds – is totally puzzling and will never be scientifically explicable. This may also be part of the worry about the 'Monster of Science' raised by Appleyard and his followers as previously mentioned. Yet there are increasing numbers of cognitive scientists and brain researchers who are gradually developing ideas which might be able to solve the problem. If one asks a psychologist of Gregory's persuasion to show how they can justify their claim that there is a 'no-go' theorem precluding us from ever understanding the mind scientifically, the only answer is to repeat the claim that it is impossible. If we built a machine with all the detailed connectivity etc. of a human brain, would we be able to say whether or not it was conscious? That seems to be a scientific question, and one I shall explore in detail later. For now, it would seem that science has a long way to go towards consciousness – and some of us are trying to take a few of what we hope will be the right steps. At this point, consciousness need not be outside the boundaries of science.

Let us turn to what appears to be the most difficult question of all – Why is there something rather than nothing? This is a form of the first and second questions about the creation of the Universe and why that particular TOE in terms of which it is explained. Indeed if the answer to the first of these questions is supplied by the chosen TOE, the question of whether there is something or nothing reduces nearly to the second question itself, since it becomes: 'what compulsion is there to choose any TOE with which to create the Universe?' However there is a difference between 'why that TOE' and 'why any TOE instead of none?' It could be that an answer to the 'why that' question also gives an answer to the 'why any' question, but it need not do so. We might, for instance, choose some criterion of elegance or simplicity (often used by scientists) to pick out that TOE from all candidate ones, but that would not help us justify why any TOE should be chosen in the first place.

It is difficult to accept that the Universe merely happens to exist. It might arguably be claimed that the answer to 'why any TOE' is the most important of all for ourselves, for it sets out the total scheme of things. In spite of being the most remote from our own daily existence, the response could well determine how we order our own lives. That is not to mean that we would not still go about our daily tasks, since at the end of the day life must go on and short of being born with silver spoons in our mouths the quality of life will be determined by our own efforts. But the answer may help us realize who or what is the final arbiter in our existence. Some such may become apparent if we could learn of the reason for all existence. The implications of the failure to give an answer are also clear – we will have to look to non-scientific sources, if any satisfactory ones exist, to have guidance in making crucial decisions in our lives.

Finally, then, can there be a scientific explanation of existence? At first glance this seems impossible. To start with one would need to consider the set of candidate theories of everything which could in some manner 'explain themselves'. For only such theories would be self-sustaining; any other type of TOE would have to be posited *ab initio*, so destroying any chance of justifying its existence. The form of such a TOE would require that its axioms or assumptions can be deducible from themselves by use of logic. As noted nearly a quarter of a century ago by an American physicist, there are no such theories of everything: 'In any event, attempts so far to derive an adequate physical theory from considerations of logic alone have not met with success. If success were achieved, I suggest that we would find that the theory was not free from assumptions which themselves need to be explained.' The answer to our question at the beginning of this paragraph therefore seems to be a resounding 'no'.

All may not be lost, however. We are glibly juggling with concepts like 'the set of all candidate theories of everything' as if they were well defined. The danger of getting into paradoxes with such ideas is well known since the time of the British philosopher Bertrand Russell at the turn of the century and even before. Moreover we are treating the problem as if some supernatural power had a list of TOEs and was scanning through them and ticking them off for choice. Why should the Universe be at all like that? It is necessary to explore more fully the nature of science, and of the Universe, before we can come to any serious conclusions. We will find that existence may be far more subtle than our above discussions would indicate.

Chapter Three

THE OPTIMISTS

The explosion of science

We are living in a period of dramatic increase in the power of science, which started about three hundred years ago with the Scientific Revolution. This in its turn was partly built upon painstaking work of earlier thinkers going back to the classical Greek period beginning in about 600 BC. Moreover, one of the important factors allowing scientists to begin to communicate freely and effectively with each other was the invention of the printing press in the fifteenth century by Gutenberg and others. Before 1500 Europe was producing books at a rate of no more than about a thousand new titles a year; by 1950 this had accelerated sharply to 120,000 titles a year. By the mid-1960s it took only three months of world production to produce what had originally taken a whole century. With the advent of desk-top publishing, the fax machine and electronic mail even more rapid styles of communication are upon us. There are now electronic networks of thousands of participants in which scientific papers are exchanged in milliseconds. The slowest part of the chain is now the human being!

Over the last three hundred years the startling thing about science is that it has doubled every fifteen or so years when measured by numbers of scientific journals, by the number of scientific publications, by scientific manpower or similar quantifiers (though not necessarily by the number of important scientific discoveries). The doubling itself is no surprise, since the human population on Earth is itself doubling every fifty or so years. But it is the rapid time scale of the doubling that is so surprising. This fact has indicated that there must come a time when this expansion of science is forced to slow down – certainly before all the men, women and children on earth have become scientists! Be that as it may, the new modes of communication now being used, and further ones being envisaged, suggest that expansion in science, measured not by the increasing number of scientists but by the increasing amount of communication between scientists, can still grow at an ever faster rate. This will emphasize even more the immediacy and modernity of science. As it is, in each 'doubling period' of the number of scientists as many come into being as in all previous time. With a doubling period of the number of scientists of fifteen years and a scientific working life of forty-five years it can be shown by simple arithmetic that there are seven scientists alive today for every eight there have ever been. In other words, most of the scientists who have ever lived are alive today.

Along with the explosive growth of science has come 'Big Science'. Large experiments performed with the enormously costly high-energy particle accelerators, such as the Large Electron Positron (LEP) machine at the Conseil Européen pour la Recherche Nucléaire (CERN) in Geneva, may require the collaboration of over a hundred highly qualified scientists. To write down all their names and affiliations on a scientific paper would take a whole densely printed page of print. The ever-increasing cost of such machines, or of big telescopes, is one of the limitations to the further explosion of science. For example, American scientists are experiencing great trouble in obtaining funding for the next big particle physics accelerator, the Superconducting Super Collider (SSC). It has been suggested by numbers of observers that the SSC will be the last such big machine. But the prophets of doom have been saying this about almost every new particle accelerator built since the 1950s. I have much faith in the ingenuity of physicists to develop new, more efficient and less costly machines to probe a long way further both within the nucleus and out into the heavens; we have not yet seen the end of Big Science.

What about new advances in scientific understanding? Have these also grown at a similar rate to the number of scientists? It is difficult to assess this, because understanding cannot be equated to the number of scientific publications. One way to assess the structure of scientific knowledge is in terms of the fundamental theories upon which science is based at any one time. In those terms there has been a clear progression towards unification and simplification with theories being developed and then being merged with others. Thus electricity and magnetism were merged in 1864 by the British physicist James Clark Maxwell into the theory of electromagnetism (which led to the development of the important media of radio and television for long-distance communication). This was then united with radioactivity in about 1974. Albert Einstein replaced Newton's laws of motion for particles by his Special Relativity in 1905 and united it with gravitation by his powerful General Theory of Gravitation in 1915. Newton's laws were also replaced, for very small (atomic-sized) particles, by quantum mechanics in 1926. Such theory is at the basis of solid-state devices in electronic chips which are now ubiquitous in industry and commerce. It was conjectured in 1982 that all the forces were united in what are called 'superstrings'. The evidence on this latter unification is not yet available but it is proving promising to many scientists working in the field. The process of unification and simplification will be described in more detail in Part Two of this book.

These giant steps have only been taken following many other smaller ones made painstakingly by innumerable scientists. One might add that the growth in total scientific output mentioned above might only be expected to lead to a much slower growth in fundamental understanding. For such steps of unification, and the earlier ones of constructing the separate theories themselves, were all based on a reductionist approach, whereby the properties of complex objects are explained in terms of the properties of their constituents. If at each step there are many, many combinations of the constituents that are realized in nature, then there are expected to be many sorts of objects which can be described as built up from the constituents. This means that there is a large compression of knowledge in going from objects to their parts. But to take the next step to the parts of those parts may require a further large amount of data, in order to get enough clues to hint at the detailed properties of these further parts. Thus we conclude that large amounts of work (both data gathering and preliminary theorizing) must be done by many for important understanding to be given by few: 'Many are called but few are chosen.'

If this is the basic nature of the *modus operandi* of science – with an explosion of activity and numbers of scientists needed to do the groundwork for the few to make a breakthrough, and with an ever-increasing range of phenomena that must be covered by such groundwork, we could expect a severe limit to the distance science can go before the earth will not have enough scientists to keep up the present rate of scientific advance. We have already noted that the number of scientists doubles at a rate three times as fast as the population as a whole. The inexorable growth of science will therefore have to slow down and be more selectively used. However, the suggestion that the TOE could be just around the corner indicates science may be self-limiting: it may have just dug its own grave! The limitations on scientific growth would then no longer be important, since no 'new Einsteins' will be needed.

The scientific optimists

SCIENTISTS ARE being propelled forward on the crest of an ever bigger wave of knowledge. This was shown by the facts we described in the previous section. It is not surprising, then, that some of these scientists have become highly optimistic about understanding the whole Universe. The problem facing them in trying to achieve a complete view of everything was as described movingly in the quotation from William Blake in the previous chapter:

> To see a World in a Grain of Sand,
> And a Heaven in a Wild Flower
> Hold Infinity in the palm of your hand,
> And Eternity in an hour'.

Can scientists ever expect to achieve such completeness of vision and can science ever explain the Universe *in toto*? The band of scientific optimists claim they can. One of their number, the cosmologist Paul Davies, phrased it succinctly in a recent radio discussion:

> It really does seem that for the first time we do have at least a glimpse of a theory which unites in one descriptive scheme all the forces of nature, all of the fundamental particles, space and time and the origin of things, just in one single theory. It's the first time this has ever happened and what we have may not be the right particular description, but at least we know what such a theory would be like.

In spite of the fact that it is certainly not the first time that there have been claims of discovery of the 'theory of everything' – witness Thales' idea that 'All is water' as I mentioned in the first chapter – it is still an amazing claim to make. Especially so since if only a glimpse has been possible how can Davies say that 'we know what such a theory would be like'? Furthermore, science is continually growing and complex problems like turbulence, weather forecasting or superconductivity are gaining increasing attention. Fundamental science is also forging ahead, with enormous strides being made this century. Moreover, the growth of all this science is continually gathering momentum. How can basic science suddenly be halted dead in its tracks by the sudden discovery of a TOE? Davies is not alone in his claims. Since the mid-1970s there has been a current of opinion among scientists that fundamental science is coming to an end. On the back of the powerful unification of electromagnetism and radioactivity in 1974 were proposals for further unification of those forces with the nuclear force in the so-called 'Grand Unified Theory' or GUT for short. At about the same time a new symmetry between Matter and radiation, called 'Supersymmetry', was proposed. The concept of symmetry in nature is ever-present, as we witness in the snowflake, in the circular ripples spreading out on the surface of a pond, or in the elegance of crystals. Symmetry seems to pervade nature down to the smallest distances inside the atom. One of the important advances in understanding elementary particles in the 1950s and 1960s was the 'eightfold way', by which the increasing numbers of particles discovered in newly constructed particle accelerators were found to have properties, such as their masses or electric charges, that allowed them to be grouped together in sets of eight (indistinguishable from each other except in specially controlled ways). This approach was important in helping to make further progress in unifying the forces.

The latest model of all the particles of nature is based on an up-dated version of this eightfold way which, instead of octets, has triplets, doublets and singlets of underlying particles in the most favoured model. This is sometimes termed the '3–2–1' model because of the sizes of these multiplets. This might all sound like numerology, but it leads to amazingly accurate agreement with the most precise facts to date available from the particle accelerators. Nature, as far as we can see now, seems based on an underlying symmetry which states that there exist three particles which are indistinguishable from each other, as are the two of the doublet. High-energy physicists have a great sense of humour, hence the name 'quark' which they coined for these three particles ('quark' is the German for rubbish and for a kind of cream cheese although the name is actually thought to have been taken from a quotation in James Joyce's *Ulysses*: 'Three quarks for Muster Mark'). Names like up, down, strangeness, charm, truth and beauty have also been used for quark properties. However, physicists have sometimes been seen to have gone too far, as when the equivalent epithet to beauty – 'bottom' – was ruled too rude to be used in a ministerial speech opening an international conference some years ago! This 3–2–1 model fits all the facts available so well that scientists are becoming worried. There is no chink in the theory's armour, no loophole through which a deeper underlying theory can be glimpsed which would allow the 3–2–1 model to be explained. It seems strange that a scientific theory can be regarded as being 'too good', when that is what science appears to be constantly striving for. Yet that is the present feeling and it gives a clear indication among working scientists that no theory can be perfect, since otherwise it would contain inexplicable components.

The new supersymmetry mentioned earlier was important in the mid-1970s because it gave rise to hope that gravity could be combined with the other forces, and this reconciled version was called 'supergravity'. For a number of reasons this hope was not realized, one of the main ones being that supergravity just could not contain the 3–2–1 symmetry of the other forces. But many scientists in the field scented victory not far away. The advent of superstrings in 1982, which was seen to contain supergravity and very likely the 3–2–1 model, now made them feel that the holy grail of the TOE was, if not here, then just around the corner. It had been 'glimpsed', or so it was claimed.

The seemingly compelling reason for supposing that the end of fundamental science was in sight is simple enough. If all of the forces of nature – electromagnetism, radioactivity, the nuclear force and gravity – had all been finally united into one superforce, why would there have been any need for further underlying force or forces behind that super theory? The complexity of nature seen around us would have been reduced, first to a set of forces operating in quite complex ways between chunks of Matter. Then these forces would have been united to give a unified force acting between a special form of Matter (represented by quarks and similar particles).

Finally this special Matter would itself have been united with the unified force by supersymmetry, so that all the material Universe, both Matter and radiation, was one unified quantity. This would interact with itself in a way dictated by the new theory of superstrings. There would be no need for a deeper underlying theory of Matter, and some scientists even claimed that there would be no room for anything else. The new super theory would be so tightly constrained as to admit nothing else. No other theory, it was suggested, could do the same trick of uniting all the known elements of the material Universe with their known characteristics. This last feature is particularly important, since the problem of uniting gravity with quantum mechanics – the development of a theory of quantum gravity – had been unsolved since about 1930. It proved so difficult a problem that, at most, one theory was expected to be a candidate. Initially supergravity, but then the superstring theory, was hailed as that candidate. This explains why Stephen Hawking could write about possible choices of the fundamental laws by a deity as:

> He would, of course, still have had the freedom to choose the laws that the Universe obeyed. This, however, may not really have been all that much of a choice; there may well be only one, or a small number, of complete unified theories, such as the heterotic string theory, that are self-consistent and allow the existence of structures as complicated as human beings who can investigate the laws of the Universe and ask about the nature of God.

It was also realized in the 1970s, and developed more fully in the 1980s, that the creation of the Universe might be looked at as a fluctuation of the vacuum. This is difficult to envisage, but one way is to think of a drop of rain suddenly appearing in a clear sky (the vacuum being the sky, the Universe the drop of rain). The dynamical framework to achieve this was called 'quantum mechanics', which had replaced the certainty of Newton's mechanics by uncertainty. Previously, particles had appeared as solid billiard balls; now they became shadowy and mysterious, without the certainty of being at any one place at a particular time. They also became prone to uncertainties in their energy or speed, and even their very existence was liable to fluctuation. But if a particle could appear or disappear by chance in quantum mechanics, so could a Universe. That might be regarded as a very large fluctuation indeed, but then that depends on the scale by which one is viewing things, and the right scale need not be so very large after all.

To explain the Universe in this way as a quantum fluctuation of the vacuum, a Universal drop suddenly appearing in the clear sky of nothing, it would be necessary to explain also the simultaneous creation of the space and time we experience in the

Universe. For that, one would require a theory in which quantum theory and gravity are properly combined. Our space and time would then suddenly appear, with the Matter of the Universe, in the primordial drop. The advent of superstring theory has given hope that such a unified theory is viable and that quantum creation of the observed Universe from nothing is scientifically feasible.

That is the brief scientific background to the emergence of optimists in the last two decades. Let us clarify what it means when it is claimed that the end of fundamental science is in sight. It is very different from saying that all science is ending, since we have already noted the enormous range of complexity in nature at all levels above the so-called fundamental. It is not these complex questions which are being answered. Fundamental science is only a part of science as a whole. Moreover, it has also left entirely to one side the problem of the mental world. How is consciousness to be explained in this TOE?

This neglect of complexity and of the nature of consciousness, as well as what some call the display of arrogance by the scientific optimists claiming to have come so quickly to the end of fundamental science, has led to strong opposition to claims that the TOE is in sight. I feel that we can discount the viewpoint that the world is too complex ever to be understood in its entirety. The existence of fundamental laws of force and of fundamental particles with elegant symmetries (like the 3–2–1 symmetry described earlier) have to be accepted even by scientists who claim that complexity is the future road for science. That there are many difficulties in going from the fundamental level at which the 3–2–1 symmetry becomes evident up to the level of weather, say, is not to be gainsayed. But there is almost no chance of new discoveries about the weather changing our understanding at the more fundamental level. Progress in comprehending the 3–2–1 theories can only really be made by probing the nature of Matter to ever shorter distances and higher energies than we have done so far. That is the reductionist route.

My criticism of the optimism over the TOE is that a TOE involving superstrings, or any other theory in which Matter alone is involved, has not now, and may never in the future, explain consciousness. The nature of the Mind has perplexed humanity over the millennia, but no advance on the problem has been made by the claim that superstrings, or any other theory, is the TOE. It is a very long way indeed from the fundamental particles of the superstring to the conscious brain. The brain works best at about room temperature but superstrings are not expected to be observable until unimaginably hotter. It is not an easy problem to bridge that gap between brain and superstring, bedevilled as it is by widespread ignorance about the brain. That somebody can write a book today claiming that 'research shows that many of us use barely 1 per cent of our brain's capacity' and that 'you could be trained to achieve a significant increase in your mental powers simply by accessing some of the massive intellect you always felt you had' indicates the enormous lack of understanding of the brain's function (as well as the credulity, if not of the author, at least of his readers). I have already quoted from the psychologist Richard Gregory, who is not alone among his scientific colleagues in claiming that consciousness itself is not explicable from brain activity. I will present arguments in Part Three supporting a purely material theory of Mind, which will later be seen to have serious implications for the way we should understand the Universe and also for how we can see ourselves in it. In particular it will make a TOE and the existence of God far less likely.

There is another telling argument against a TOE ever being found. It is apparent that any such TOE would be expressed in mathematical form. Think how antiquated it would look to mathematical physicists thousands of years hence. Mathematics would most likely have advanced enormously in that time. Future mathematical structures and concepts will make those of today appear trivial in the same way that our present mathematics diminishes that, for example, before the invention of the calculus. Vast new mathematical disciplines will have been developed. None of these would have been used in any TOE which is developed over the next few years. Up till now physical theories have been expressed in terms of the latest mathematics. This leads me to suggest that the unending future development of mathematics will help produce models for ever more sophisticated versions of the material world. With no end to mathematics in sight there should be no end to its usefulness in explaining ever deeper levels of reality.

From another point of view any claim that the end of fundamental science is in sight could be more appropriately regarded as pessimism than optimism. There would no longer be the thrill of the voyage of exploration, of forging ahead into the intellectual unknown. The frissons of the scientific race would no longer be felt. There would be no struggle to be the new Einstein among the cleverer young men and women, nor stimulus for them to compete for the glittering prizes; the tree of knowledge would have been stripped bare of its fundamental fruit. Mount Everything would have been climbed for the first and last time. After that one could just take the ski lift to the top – and where would be the thrill in that?

This seems a depressing picture. But other answers, especially those from the past, should be considered to help guide us on our way, and see how to fit the possible TOE into a broader framework.

Answers from the past

It may be helpful to look at the past, not only in terms of science but also of religion, to see if the scientific optimists – claiming that all would soon be explained – can gain any justification for their dramatic claims. Did scientists of the past expect such a possibility to occur? How could the world's religions relate to this fantastic possibility? At the same time we should also assess, in their own right, answers from religion to the 'impossible questions'. This is a vast subject, filling many libraries. I do not propose to give a potted history of the world's science and religions, but just to dip into those parts which I feel are illuminating. My excuse for an apparently sketchy treatment is that I want to explicate present ideas, not past ones.

It is particularly appropriate to indicate why this apparent neglect of past ideas is justifiable. It is because we are using the intellectual tradition which builds on past ideas. Those ideas that are good are kept, at least in as far as they are helpful in solving new problems. The advance of science has been achieved precisely through such an approach. Due to the fact that most scientists who ever lived are alive today – the 'immediacy effect' which I described earlier in this chapter – most of the effective ideas now are relatively recent, at least in the material sciences. This is also becoming true in the life sciences. Unlike books in the arts (which do not date) all scientific books have a built-in obsolescence. Shakespeare does not date (I took his collected works with me on a visit to CERN, the particle-physics laboratory in Geneva, but only had the latest scientific texts with me). That the impossible problems previously mentioned have been with us since recorded time does not mean that proposed solutions by the Ancients must all be considered, since these proposals would have been made without the benefit of modern knowledge. Even so, there are attitudes and styles, as well as content, to learn from the past. History of science is a specialized subject, but has a legitimate place, for example, in a science degree course.

The natural philosophers of ancient Greece were renowned for their 'theories of everything'. Thales' choice of 'water', followed by the 'air' of Anaximines and of an unknown, ageless, underlying substratum by Anaximander, were noted as early candidates. The atoms of Democritus gave a modern-looking approach to early Greek ideas while the Ionian Anaxagoras believed that Matter was infinitely divisible and that Mind was a separate substance which entered into the composition of living things, and was a source of all motion. It was Pythagoras who said that 'all things are numbers' and from him can be traced the modern idea that mathematics is the language of fundamental scientific theories. Indeed the level of mathematics required now to understand the theory of superstrings is very high and one needs several mathematical courses to understand it. However, I suppose that the task of comprehension is no harder, relatively speaking, than it was in Pythagoras' day to understand his famous geometrical theorem learnt still (I hope) by every schoolboy and girl.

The later Greek philosophers' ideas, such as those of Plato, developed partly from the belief that mathematics was the chief source of eternal truths about exact or ideal entities; hence Plato's forms, which were claimed to be the true reality behind mere appearance. For example, the form of justice or of beauty was supposed to exist independently of and prior to just actions or beautiful people. Later philosophers also had a blend of moral aspiration with rationalism which even underlies this book and the attitudes of many thinkers today.

But we must recognize that both Plato and the other important Greek philosopher Aristotle (Plato's star pupil) were both opposed to the much more modern-looking materialism of Democritus who claimed that everything – life, order, mind, civilization, art, etc. – can be reduced to the movements of particles of Matter controlled by certain laws.

The ideas of Pythagoras took strongest root in the work of Isaac Newton. The latter's three laws of motion and his 'inverse square' force law of gravity allowed the development of a framework of mathematical equations leading to explanations of a range of natural phenomena and predictions of others. Mathematically expressed theories and their testing by experiment became normal for scientific development, and modern physics and astronomy use sophisticated mathematical ideas – matrices and vectors, differential and integral equations, infinite dimensional spaces of functions, group theory, topology, and other branches of mathematics to give an ever-increasing comprehension of the natural world. Mathematics is rightly called the queen of the sciences. The scientific optimists are correct, then, in their expectation that their TOE would be expressed in mathematical form.

The neglect of the problem presented by Mind is to be expected from physical scientists. The detailed study of brain and mind is still in its relative infancy. It was only in 1865 that two German surgeons experimented on the brains of dogs to show how electrical stimulation of parts of the brain led to various movements. The discovery that different areas of the brain perform different functions has not been developed very rapidly. Only now are hints becoming available of how connections between various cortical areas and midbrain regions may lead to consciousness. It was at the turn of the century that Freud began developing his theory of the subconscious mind. All this knowledge of the brain, albeit of a very complicated system, is at a far more superficial level than that of our understanding of Matter. The 'mental' scientific revolution began more than two hundred or so years after the 'matter' scientific revolution. The former has a great deal of leeway to catch up on the latter. The reasons for this sad state of affairs (that we know far, far less about our 'selves' than about our bodies) are to be found both in the general attitude to the mind by the philosophers and theologians of the past and by the intrinsic problems it presents to the investigator.

The attitude of the world's religions to the scientific optimists is obvious, since a religious believer already has his own answers to the impossible questions, provided for him by his religion. Not that such answers need be understood. Thus Christianity would say (in Wisdom 9: 9–17) 'the reasoning of mortals is worthless', and the Early Christian thinker St Anselm stated: 'Unless I first believe I shall not understand.' The main items of faith of a theist would appear to be that the soul is non-material and can have direct relation with God. These underlying assumptions are a clear thread in the Bible, starting with the very first chapter of Genesis. They are also strong in the writings of modern theologians, such as those of the physicist-turned-vicar, the Revd John Polkinghorne, President of Queens' College, Cambridge: 'A second reason for taking the Transcendent seriously begins with my own experience of prayer and worship. However fitful and elusive my experience of God may be, it is not to be denied.'

More generally, any theistic religion must use its God as the source of extra-material existence. The material world may become ever better understood by physical scientists but the Mind or soul is claimed to be impenetrable by such approaches. Brain and Mind cannot be equated; the old adage 'What is Mind? never

Matter; what is Matter? never Mind' is at the root of the theistic approach to life. This is at the basis of Christianity, Islam and Judaism. Hinduism also has a theistic component. On the other hand, one branch of Buddhism (the Theravada line) has not related itself to any theistic approach, while another (the Mahayana branch) has the idea of Gautama as the incarnation of the 'Buddha-spirit'. More generally, any belief in reincarnation would seem to require an intangible, immaterial soul or self, which may be claimed to persist, for example, through subsequent lives as one rids oneself of one's karma (the guilt acquired due to misdoings in this or earlier lives).

The response of a theist to the claims that superstrings (or whatever) is the TOE, would most likely be to raise the 'impossible questions'. To the first, 'How did the Universe begin', the scientific optimist could respond that the Universe was created from nothing by a quantum fluctuation. The theist need not be impressed, since there would be no answer to his question 'Where did quantum mechanics come from?' To the second, 'Why that TOE?', the TOE enthusiast could only say, 'Because it fits all the facts', which may not be very satisfying to the theist (but appears to be on a par, in principle, with the theist's belief in his or her own theology and its answers). To the third question, 'How can the Mind be explained?', there may be the scientific claim that consciousness could arise in terms of the complex functioning of the brain (that explanation being accompanied with a lot of handwaving but very little in the way of detail), to which the theist could rightly reply, 'I can't believe you until you have put in the details. Since you have not yet done so, I therefore do not need to accept that you could do it.' The final question, 'Why is there something rather than nothing?', would leave the scientific optimist either completely stumped or appealing to the anthropic principle. That runs along the lines of 'We exist, therefore the Universe must be such as to allow us to exist and discuss its existence.' That seems a rather weak argument. We are indeed here: but to argue in this way seems like saying 'We are here because we are here.' Theists would feel correct in claiming that God gives a better reason for existence than such an approach.

The theist does have a delicate tightrope to balance, if challenged, as follows. Suppose that an artificial brain were made, duplicating exactly his/her own. Would that brain be conscious? This question only highlights the problem of 'God of the gaps': how can God leave unchanged the physical laws yet allow our (immaterial) minds to influence our physical bodies? There is no realistic mechanism which has ever been suggested to allow Mind to interact with Matter. It is possible to answer by regarding Mind and Matter as two aspects of an underlying entity. But the nature of this underlying entity seems to be highly mysterious. Moreover, the persistence of Mind after death is then a problem, because the nerve-cell activity, the material basis of Mind, will have disappeared. How can the mental aspect persist? What can it be 'activity' of, if no physical energy is involved? If Mind is absorbed as 'information' in God (as some theologians claim) there is then the difficulty of describing this 'information' which is contained in non-material entities such as the immortal mind or soul. For this entity, being outside space and time, appears impossible to quantify. But the information you and I possess in our brains has physical manifestation in terms of memory. If that memory disappears at death, so does our information. We cannot therefore retain any of our personal identity after death. This problem has never been solved and I do not see any way of so doing.

This is an appropriate point at which to raise a question which is at the basis of the above problem. How can one meaningfully discuss immaterial quantities? In describing something to somebody else I may attempt to invoke sense in, say, my description of a beautiful sunset I had seen the previous evening if I use adjectives like 'wondrous', 'glowing', etc. But I am not sure I achieve more than a general evocation of the scene and of my own response to it. If I were to give the numerical values of the wavelengths of the colours at each point of the scene then I would give a far more precise description, one which would be helpful in evocating the scene for somebody else. It seems that this is an example that supports the rule: only if one can quantify what one describes can one impart information that is verifiable by others.

Almost by definition, 'immaterial' is non-quantifiable (what is the weight, length or colour of the Mind, or of God?). So how can an immaterial entity be described clearly to others (and hence to oneself)?

We can summarize therefore that the science of the past has naturally led to the search for the TOE going on today, with the form of the TOE being embodied as a set of mathematical equations. The theist is likely to criticize any such TOE for its lack of ability to answer at least two if not three of the impossible questions; any theology worth its name should be able to answer them all. However the religious answer does not appear very clear, nor indeed very successful.

Pitfalls in prediction

THE SCIENTIFIC optimists can be attacked on grounds of pride in claiming far more than they should. They should take note of numerous examples of predictions for the future which were shot down in flames within a short space of time. Lord Rutherford, the 'father of the nucleus' (so-called for his discovering the existence, in each atom, of a nucleus in which nearly all its mass is concentrated) said in 1932: 'We will never crack the power of the nucleus.' Yet only a few years later the first atomic bomb was exploded in New Mexico and shortly thereafter two atomic bombs were dropped with devastating effects on Japanese cities. Then there was the British Postmaster-General who, when advised of the discovery of the telephone by Alexander Graham Bell, was reported to have replied: 'We have no need of the telephone; we have plenty of messenger boys.' At the turn of the century a distinguished scientist stated that there was nothing left to invent. At about the same time, another distinguished scientist claimed that: 'Science has come to an end. It is now only a question of filling in further decimal places.'

If we take the ups and downs of the predictions of 'theories of everything', from the beginning of rationalism in ancient Greece up until today, I do not think it very likely odds that this time the TOE will be the final one. In any case the TOE will still have to answer most of the 'impossible questions' set out previously. In the next chapter I will begin to put together my answers to those questions.

Chapter Four SCIENTIFIC SOLUTIONS

Creation from nothing

IT IS DIFFICULT for us to conceive of something arising from nothing. In the first verse of Genesis it is written: 'In the beginning God created the heaven and the earth.' However, there is no detailed account of what materials God used in the process nor, more importantly, where such stuff as was needed came from. Were they created out of nothing, or changed from something already existing? Yet we might accept the biblical account as written because we subconsciously regard God as having all the requirements already to hand. We might even say that God is the Great Conservationist, since he allows us to see the books balanced in the greatest creation act of all time. He did not cause pollution, but simply recycled things. Where did the energy and matter come from to make the Universe? It was God's – part of himself, so to speak. Where was the space and time in which this creative act occurred? Again, it could be said to be part of God's 'person'.

The recent developments in science have allowed us to dream up scenarios for this generative act which could be regarded as requiring less and less of a truly active process on God's part. Some scientists are even claiming, as I briefly described earlier, that God's presence or activity in that initial stage has dwindled into insignificance. It corresponds to what we can describe as obtaining order out of chaos. However, before we consider the seemingly miraculous emergence of structure from formlessness, we need to make crystal clear exactly how much of a structure we must erect before we can have anything arising from the void.

Energy in the Universe is stored either in the gravitational attraction between galaxies or in the kinetic energy of their motion (the 'mc^2' energy of Einstein). Gravity, because it is attractive, can help in reducing the energy balance problem when Matter is being created. The gravitational attraction between massive bodies means that gravitational energy is negative (it requires energy from outside to separate massive bodies) and so could cancel out the kinetic energy of the bodies. Suppose we liken the force of gravity between two particles to an elastic band stretched between them and release the particles when they have been stretched apart. They would move towards each other, propelled by the energy stored in the elastic. This elastic (gravitational) energy is turned into energy of movement (kinetic energy). In gaining kinetic energy they will gain mass (by energy = mc^2). They might then become so heavy they could break up into smaller particles. In this way new Matter would be created, coming not from the void but from the transmutation of gravitational energy into Matter.

A back-of-an-envelope calculation can be done to show that if all the negative gravitational energy of the galaxies distributed about the Universe were added to the 'mc' energy of those galaxies the total would be zero. This is an amazing fact with no real explanation but an important implication. The total energy 'bank account' would be empty. Therefore it need not cost anything *in toto* to create any new Matter; the positive energy cost of creating Matter would be paid for out of an increasing overdraft run up on the gravitational energy part of the account. Experiment has indicated to us that the total energy, gravitational plus kinetic, is always kept the same. If there is to be a Big Crunch, when all the galaxies cease to fly apart but fall back towards each other and finally coalesce, we would expect the two energy accounts (the negative gravitational and the positive kinetic energies) to balance each other out exactly. There would then be nothing left at all when the Big Crunch was over. The total energy account would be closed. The Universe would no longer exist as an object with energy of any kind. There would be nothing to see from a table in Douglas Adams' 'Restaurant at the end of the Universe'. All would be dark.

There are several further aspects which need clarification in this scenario. First we have to discover how the things we see around us – planets, stars and galaxies in the heavens – can have taken the place of a mere structureless mess. That may not be so difficult, since it is now becoming clear that order can arise from disorder. This happens in ourselves, through the growth of our bodies by absorbing energy from our environment and turning it into specific tissues. We can begin to understand how this might also occur for matter at various scales, even down to that of elementary particles. The concept here is that of the bootstrap, made famous by the well-known fictional tall-storyteller, the German Baron Munchausen. He claimed that in order to save himself from sinking into a bog at one point in his career he leant down, and pulling hard upwards on his shoe laces was able to 'pull himself up by his bootstraps'. It is through elementary particles interacting with each other so as to create only a special subset of themselves (and no others) that such a bootstrap process is suggested to occur. It is as if these special particles are able to pull themselves up by their own bootstraps (which in their case are the forces between them) to create themselves from nothing, as Baron Munchausen saved himself without visible means of support. Combined with getting energy in the Matter energy bank account from the gravitational energy bank account overdraft (remember that the total of Matter and gravitational energy bank accounts is always an unchanging amount, which we are taking as zero), this bootstrapping has been proposed as a scientifically respectable scenario for creating a highly specialized Universe from nothing. The second question to resolve is that of the time of the cosmic generative act. Note that we are speaking of time as if it were already present as an eternal clock determining when and how fast things happen. We will consider how to avoid this rather *ad hoc* feature in the next section. Given that there is such an independent time, how and when could the Universe suddenly appear?

The 'how' part of this question appears to be a little easier to answer than the 'when'. To answer 'when' it is not enough to use the ideas of classical physics, in which particles are at certain places with certain velocities at a given time. Such certainty would not allow particles suddenly to appear or disappear, and we therefore have to turn to quantum mechanics for the creation of particles at random to be possible. Particle creation has been observed in various of the particle accelerators, such as the one at Geneva. So we know it can occur as a random event, as long as energy is provided to give the created particles their mass.

It is proposed that the Universe came into existence as a 'quantum fluctuation' of the vacuum. It just 'popped into being', so to speak. I noted earlier that there would still be a large gravitational energy debt to repay at some possible future Big Crunch. Now a severe problem arises about the time at which the Universe appeared. For we have assumed that time has been rolling inexorably and uneventfully on for ever, no instant being singled out from any other. So it does not appear possible to choose any special time from any other when the cosmic creation act would have taken place. All we can say is that there would be a certain probability of such a creative process occurring in any interval of time. If our Universe was created in one time interval then other Universes could be created in other time intervals. Since in this approach time has been going on for ever (it is not the time that is being created but the Universe), then an infinite number of Universes are expected to have been created. Such Universes could get in each other's way. However there is no hint of any collisions occurring. The distribution of matter in our Universe is beautifully smooth and homogeneous. No sign of shear or strain which would have signalled such collisions is apparent. So this idea of the Universe as a 'quantum fluctuation' in an ever-rolling stream of time has severe difficulties. In any case, time (and space) have been taken for granted, so while we have seen how Universes might be created without destroying the overall energy balance we have not been led to building space and time themselves. That is not enough for modern thought.

No beginning

THE NEXT STEP forward is to attempt to construct a theory of space and time in which these quantities are so closely interlinked with Matter that creation of the latter leads inexorably to simultaneous formation of the former. This requires that we abandon the notions of transcendent space and time which are not involved in the hurly-burly of dynamical existence. Sir Isaac Newton believed in a single space and independent time through which the earth and planets moved. It was Albert Einstein in 1905 who showed, in his Special Theory of Relativity, that space and time are only relative to the way they are observed. His famous dictum 'moving clocks go slow', which will be explained later, clearly indicates that the rate of passage of a clock is dependent on its motion. It was also shown by Einstein that if a ruler used to make spatial measurements moves parallel to an identical stationary ruler the spatial measure of the moving one contracts. But space and time are still external constructs in special relativity, in spite of their being relative to the movement of anyone measuring them. Even in a Universe empty of Matter such space and time would still exist. It was not until Einstein developed his General Theory of Relativity in 1915 that it became possible to tie space–time inexorably to Matter. This was achieved by using Matter as the source of the space–time structure in which the Matter was moving. This structure is particularly defined in terms of the distances between pairs of events. Such sets of distances effectively give a 'triangulation' of the space–time in the same way that a surveyor's map can do for a building site. If the Matter distribution is known, the corresponding surveyor's triangulation map of distances can be calculated, and there is excellent agreement with the real thing when the two can be compared, as it can for planetary motion.

In Einstein's General Theory of Relativity, time is no longer well defined but can, for example, be defined by clocks moving at a variety of speeds, none with preferred status. That is to be expected in the same way as it was in the special relativity theory (where moving clocks are seen by stationary observers to go slow). However, the situation is now more crucial because it is hard to discuss the evolution of a created Universe when there is no obvious notion of the time to be used to measure development alongside. Let us suppose that such a choice can somehow be made (and there is no reason in principle why it cannot). We then meet the far more difficult problem of combining Einstein's theory and the fluctuations produced by quantum mechanics in order to generate both space, time and matter in the cosmic creation. In the process we would hope to have overcome the problem we had at the end of the last section, of how to avoid the creation of a plethora of Universes (which seemed contrary to the evidence), as well as to go beyond that to fabricate also the space and time of the cosmos.

There are numerous problems in consummating the marriage of quantum mechanics and gravity to produce quantum gravity. These barriers were mentioned in Chapter 3 and will be discussed more fully later. Whatever form of theory will finally result, it is not expected to have time associated with it intrinsically. Any notion of time must arise from the way the Universe is modified so as to describe this change in the most appropriate manner. Whatever the results of that investigation might be, they lead to a new problem called the 'initial value problem': in what state would the Universe have been when it was created? It would seem that the use of the word 'created' would imply creation 'in time'. However we have already noted that we expect to discover candidate 'times' by more careful study of the system. We are therefore led to the problem of singling out one special state of the Universe from all other possibilities. That means we have to discover reasons for there being so many galaxies of certain sorts and so many particles with certain properties, and why they are not otherwise.

Let us try to avoid the problem of discovering this 'special state' by dispensing with quantum mechanics. If we just use Einstein's general theory it leads to equations in which Matter and gravity develop their activity in time. But to find out what that activity is at a particular time we have to specify what it was at some earlier time. We may push that initialization back ever further into the past but we still have to face up to specifying what activity there was at some earlier time. We cannot seem to avoid the initial value problem: what was the initial state of Matter in the Universe just as it was being created?

One way out would be to push time back ever further and further, so as to effectively claim that the Universe 'always' existed. That is an idea we will return to shortly, but note that it cannot be made compatible with Einstein's equations, the laws of classical physics for Matter, and the observed facts of the speeding of the galaxies away from each other. As we noted earlier, if the galaxies are followed back in time it is clear that they were condensed to a single point about fifteen billion years ago. Earlier than that, the Matter of the Universe disappears, and therefore the laws of classical physics break down, since they just do not describe the creation or annihilation of Matter. It would be necessary after all to bring in quantum mechanics to do that and then, as I have already said, time can no longer be explicit in our description.

An alternative approach is to suppose that the Universe is in a cyclic motion, expanding-contracting-expanding-contracting and so on. This meets with the difficulty of explaining the bounce, which again would seem not to be possible by classical physics (including general relativity) alone. Thus we seem to be forced into facing up to how quantum gravity would describe the beginning of the Universe, and more especially why the state of the emerged Universe is as it was rather than in any other state.

There have been various attempts to discover a natural specification of a unique quantum mechanical state of the Universe, but none have yet been completely agreed on. Nor have any led to experimental justification. One such attempt, suggested by Stephen Hawking and an American colleague James Hartle, involves considering all possible four-dimensional space–times into which the three-dimensional space could evolve. The model uses an essential trick of making time purely imaginary (multiplying it by $\sqrt{-1}$). This corresponds to making space and time similar; in imaginary time we can go round in circles exactly as one can in space. Difficulties of travelling back in time disappear. One way this can be visualized is if we imagine ourselves living on the circumference of a circle. Addition of imaginary time would lead to us extending the circle to a two-dimensional surface like a soap bubble blown from a circular metal ring as support. The four-dimensional space–times that we live in are like the surface of the soap bubble; their three-dimensional boundaries (from which they have evolved) are like the supporting ring. This approach leads to a unique state and interesting features of simplified models.

Another model uses conditions on the state for three-dimensional spaces which become infinitely small, like black holes (because the probability of any possible three-dimensional geometry, including a nasty one like a black hole, is potentially included in the quantum state of the Universe). This approach also leads to intriguing features of such simplified models. While these ideas are still only scenarios for still-to-be-developed working models, the possibilities indicate that some type of constraint might be discovered to give a state of the Universe consistent with its later development as we can observe it today. Can we conclude then, that such an approach, together with a TOE used to calculate the state of the Universe, could lead to the end of theoretical physics?

Why that TOE?

THAT THE END of theoretical physics might be in sight has been the track taken recently by a number of scientists and I described some of their pronouncements earlier. The claim was bolstered by the apparently rapid advance towards a theory of everything related to the discovery firstly of supergravity and later of superstrings as a solution to the problem of construction of a good theory of quantum gravity. The feeling of euphoria this rapid progress generated was compounded by the imaginative scenarios for the quantum creation of the Universe which I described briefly in the previous section.

Having apparently got so close to the 'holy grail', surely such high-spirited optimism might be well founded? To recognize the danger of being swept away by the flood of words on the subject appearing recently in the popular press, including the 'popular' (but not necessarily understood) writings of some of the protagonists, we must probe beneath the surface skin of these apparently attractive forms of the modern scientific vernacular. An important question to answer is what sort of theory has to be used to explain the creation of the cosmos. For that certainly would give clues as to the nature of any deeper TOE. In the previous section we concluded that a sensible scenario could possibly emerge from the quantum creation approach. But that was based on the use of the twin theories of quantum mechanics and gravity, which themselves required an underlying theory such as superstrings to unite them. We must therefore only take evidence on the nature of the TOE from attempts to use superstrings to explain the quantum creation of the Universe.

The scenarios described (in the work cited earlier of Hawking *et al.*), together with many others, are imaginative but are using an incomplete framework. They single out gravity as crucial in very extreme stages of the Universe. We should not, therefore, accept these claims about explaining the creation of the Universe as having relevance to the status of the TOE. Let us consider the status of the existence of the TOE more directly.

The question 'Why that TOE?' was earlier classified as an impossible one. It was claimed that the scientific answer to that question could be obtained by ensuring that all the predictions of the chosen TOE fitted all the possibly observable facts. That may well be as far as the scientific method can allow one to go to answer the question. There have been many candidate theories, such as Thales' 'All is water', but few now appear viable. Among these latter, one might use the anthropic argument that only this particular TOE could lead to life, other candidate theories never giving the right conditions for life to evolve as the Universe developed, and hence to any discussion of the question itself. As when this approach was raised earlier, it still appears unsatisfactory. It is true that our conscious existence would be deduced from the TOE and the resulting evolution it predicts for the Universe, but our existence and its nature is only one of many empirical constraints on the TOE. There seems to be no reason to elevate this to a religious principle.

Another popular approach is to claim that there is a selection mechanism of some sort choosing that particular TOE from the set of all candidate ones. The exact nature of that set of candidates is, of course, debatable and so also is the criterion that might be used. To avoid reducing the whole discussion to a matter of personal preference it has been suggested that there is an 'intelligent' selector at large who is making that choice in a way beyond our ken. But this comes back to some sort of God behind it all, a solution which I will try to avoid since I have already established that it provides no details. Moreover if we do believe that God was the best of all mathematicians, the level of mathematics needed to state the TOE may always be beyond our grasp.

The bootstrap was mentioned earlier as a dynamical mechanism to create order out of chaos, and in particular to produce the observed elementary particles. Could the same approach not be used to bootstrap a theory from itself, so to speak, and in that way select only the TOE (by the bootstrap mechanism) which it has been claimed we are just now dimly glimpsing? Sadly, in spite of the elegance of the idea, it does not seem to work. I said earlier that it is just impossible to derive a theory from its own premises. That is the same as saying that the theory cannot be bootstrapped.

To summarize the situation, we do not seem to be able to find any way to justify the TOE, yet feel cheated by being unable to answer the question 'Why that TOE?' We can add that we might also feel led to question the very possibility of a TOE, since the sequence of theories we have had so far in the scientific revolution has given us no strong hint of coming to an end. Let us explore that important aspect further.

At each critical step of scientific development in which a new theory is developed and justified, the theory it has displaced may, it is claimed, have been 'explained' by the better one. Thus the laws of classical mechanics (Newton's famous laws of motion), may be deduced from the underlying quantum mechanics, under the conditions in which each particle involved is so heavy that the distribution of its probability for being found at any point in space reduces to a certainty that it is found at one particular point, as for a billiard ball. The presence of a chain of theories, explaining each other in the above manner, would thus seem to be an essential feature of the world in which we live. If that sequence of theories were unending then everything, including every theory, has its cause. The chain should never end, however, since if it did the final theory of the chain would be the TOE without reason, which I mentioned above as being unsatisfactory. That would destroy the possibility of always giving a causal explanation of each theory of the chain by the subsequent one.

One might, at this point, claim that this whole edifice of an infinite chain of theories does not, itself, have a cause. Seen in this manner we do indeed have a puzzle, but the chain of theories being uncovered is not itself also a theory. All that we are able to do in science is to test any theory to its limits. Once a theory has been found wanting, a new and better theory is developed to replace it, so explaining the previous one along the lines we mentioned above. It does not seem possible to probe, in any scientific manner, the nature of the sequence of theories that we are discovering in our scientific odyssey. Indeed at each time we only have the present scientific theory (or theories) we think valid today, not the previously rejected ones. Moreover, the nature of the chain of theories can never be completely apparent to us, since in this approach there will always be an infinite number of theories still to be uncovered, each displacing the previous one. We might conjecture what is going on after a few steps, and this might help us in our infinite quest, but the nature of any all-embracing theory of theories will always be absent. For that itself would appear to have the nature of a TOE and we have to accept that such a theory is an illusion if we wish there always to be a causal explanation to any scientific theory.

Why should we put such faith in science as a way of understanding the Universe? Why not appeal to God, instead of leading ourselves into an infinite regress? One reason, already mentioned, is that only through science can we be sure of what we mean. But then we can apply this surety of meaning to the Universe as well. Only through quantification – especially through description by mathematical equations – does the Universe become explicable. That applies not only to our understanding of it but also to its own ability to cohere. It would seem that various parts of the Universe would not know how to combine together unless the future actions were quantifiable. That is a lesson we should have learnt well by now – all of the theories which have been successful since the scientific revolution began are only properly expressible in mathematics. Therefore all our theories of the Universe appear to be mathematical, with sets of axioms for certain of the fundamental quantities (particles, fields, etc.). Only in this manner would the Universe be able to evolve in a well-defined manner, governed by precise mathematical laws.

We come, then, to a picture of the Universe as an infinite set of levels. Each of these is described or governed by a well-defined set of theories, expressible in mathematical terms. The theory at each level is 'explained' by that at the level below. Newton's classical mechanics, for example, can be explained by deriving it from the underlying quantum mechanics. The latter can be explained from the more complete quantum field theory, and so on. Thus the question of this section should be 'How that TOE?', and seems to be answerable in scientific terms if 'TOE' were replaced by 'present scientific theory'.

This leads me to suggest that there is no ultimate TOE, but only a sequence of scientific theories to explain the Universe, reaching ever greater precision. Each theory displaces the one before it and answers the 'how' question about that displaced theory; the new theory will have greater explanatory power to allow it to fit the facts which were recalcitrant for the preceding discarded theory. This seems to be the only way in which causality is never violated. The Universe not only *is* mysterious but it will *always* be mysterious. There will be no hindrance to parting the veils of the temple but there will always be another veil to part.

Consciousness

THE BRAIN OF vertebrates evolved over the five-hundred-million-year span of vertebrate history. Fossil records have recorded the details of this evolution. Birds and mammals have been found to be 'higher' vertebrates because they are more highly encephalized than reptiles and other 'lower' vertebrates. It is even possible to quantify this difference in terms of the ratio of the total brain size divided by the two-thirds power of the body size (where this two-thirds power effectively reduces the volume of the animal's body to its surface area, which is the quantity needing control). This ratio, called the index of encephalization, is about the same for all reptiles, both fossil and alive today. Birds and mammals evolved independently from two different sub-classes of reptiles. Mammalian encephalization did not progress at once but stayed at a steady level of increase (with a fourfold brain expansion over lower vertebrates) from about a hundred and seventy million years ago for the subsequent hundred million years. In the last sixty-five million years there has been a larger, but still steady, evolution to the present value for mammals such as carnivores, monkeys and apes. Man fits into this, but only as a very recent newcomer, his brain reaching its present size only in the last million years, with an explosive doubling over that period.

This doubling of brain size was very likely brought about by increased survival chances granted to individuals with bigger brains. Such an evolutionary growth of the brain would be expected to have a concomitant growth of mental abilities. In particular more brain would have allowed for the emergence of stored representations of external events and environments. These stored memories would have been able to have been used ever more effectively in planning actions (by appeal to past experience) as brain size increased. Such internal representations might have been re-excited to make such forward planning more efficient, so allowing the development of imagery. Driven on by 'survival of the fittest', ever stronger mental powers would have granted the possessors ascendance over the less quick-witted. The ascendancy of Mind over Matter had begun.

Where in this evolutionary process could it be claimed that Mind and Matter should be recognized as separate, or that Mind had been attached from 'outside', so to speak? Creationists, for example, would deny the continuity of the evolutionary process. Theists would claim that Mind must have more than purely physical attributes. But it appears very hard to say 'here it was – mind occurred just here, and no earlier in the evolutionary process', or 'after this date mind became related to God'. The palaeoneurological evidence mentioned above seems strongly to support the simpler and more direct supposition that Mind developed solely because the brain became more complex and that the process of consciousness arose purely and magnificently from the subtle activity of the billions of active nerve cells in the brain, each with its complex structure but combining together to produce the scientific genius of Albert Einstein, the musical creativity of Ludwig van Beethoven, or the literary imagination of William Shakespeare. On these grounds there would be a continuum of levels of consciousness, starting with the lowest vertebrates and increasing as encephalization occurred until we reach the highest level with man.

The idea that consciousness arises solely from the activity of the brain is a simple and appealing one and has many scientific supporters, especially among neuroscientists. It also has its opponents within the ranks of such people. It has to be admitted that little progress has actually been made in tackling the problem of how consciousness arises from brain activity. Thus a distinguished neuropsychologist wrote a few years ago: 'Even the most generous reading of the clinical and experimental literature that constitutes a century's worth of neuropsychological research leads to the conclusion that the contribution has been rather modest.' This lack of effort, as well as progress, may be ascribed partly to the strong opposition by the influential behaviourist school of psychology over several decades. Scientists also have to work on problems that can be solved. As the Hungarian scientist Polya said, 'Science is the art of the possible.' The study of consciousness and Mind did not seem to be among 'the possible'. Introspection was not to be trusted, and consciousness of others or animals was claimed to be unprovable. Not until very recently could it be argued that explaining consciousness was a problem with any hope of success. The detailed neurophysiological basis of consciousness is only now being elucidated. New tools, such as those allowing measurement of the minute magnetic fields produced by brain activity or those measuring the increased energy consumption in active brain regions, have finally allowed for localization of some of the processes involved in attention, which can be regarded as the gateway to consciousness. More precise studies of the attentional deficits of patients with brain lesions has added to that understanding. The time is now becoming ripe for a concerted attack on the problem of constructing scientific models of consciousness based on brain activity. Only after extensive testing could any of these be accepted but without such attempts science will be shirking the immense challenge posed by the Mind.

The form of solution I propose here is based on my own research on the problem over the last twenty years. The basic idea itself of the nature of the Mind is simple and has already been stated (but in a less extreme and neurophysiologically grounded form) by the Associationist school of philosophers and psychologists. The concept of associationism was that ideas could be hooked together and there were certain laws of association (such as contiguity, similarity and intensity) which developed into modern learning theory. Instead of taking ideas without explanation and asking for the manner in which they can be related and developed, I wish to suggest that ideas or mental experiences themselves are in a given person composed of the set of relations between the external inputs and the memories of past events these inputs arouse in the brain of the percipient. This is what I call the

Relational Theory of Mind. It implies that the 'colour' or content of conscious experience of an event is determined not only by the inputs from that event – the visual images, the sounds, smells or touches – but also by the stored memories of similar experiences in the past. All mental content is in the relations, and a baby will have a much weaker consciousness than a mature adult because it has fewer such relations.

The Relational Theory is still only a general framework; it does not seem to lend itself to any specific tests. It must be implemented by developing a model of the tissue of the brain in which the crucial features of consciousness are apparent. That necessarily means that the mechanics of implementing the relational structure itself must be proposed and tested. Moreover the conjectured essential brain structures must lead to the inescapably unique but continuous stream of consciousness. We can attend to only one thing at a time, but our consciousness flows on without gaps when we are awake.

There are areas in the midbrain and cortex (the surface of the brain) – especially the so-called nucleus reticularis of the thalamus, which forms a continuous well-connected sheet of nerve cells around the input to nerve cells in the thalamus (being the relay station for all sensory inputs to cortex) – which seem to function as a template or mask rapidly to correlate all activity going up to the cortex or back from it. This cortical activity itself can persist over long enough periods of time, by feedback circuitry, so as to allow for the stream of consciousness we experience in our waking hours. The uniqueness of this stream may be achieved by the nucleus reticularis sheet only allowing one correlated activity at a time (which is the winner of a competition with other incoming activities played out on this sheet) as if it were composed of a set of templates which can change with time but only in a limited fashion. This control of cortical activity by mid-brain structures seems to increase with age, so that the older person has less flexibility of thought by cortical processing and is more controlled by his mid-brain activity; the patient with Alzheimer's disease has even less cortical activity, so even less conscious content. These and other aspects of the neurophysiological basis for this theory will be discussed in more detail later.

The Relational Theory of Mind gives an explanation of the apparently non-physical nature of conscious experience and a possible source of the notion of free will. For the set of relations between brain activities (both present and stored in memory of the past), the very essence of the relational theory, is not itself a physical thing such as brain activity. It is more tenuous and also made highly complex by the myriad relations that have been set up by a person's past experiences. Because there is such subtlety in the set of relations at any one time it may be difficult, if not impossible, for a person to recognize that such a relational activity is going on inside their brains. They will have the impression of a non-physical process going on – their consciousness – which has no clear rules of control. In this way, freedom of action – free will – will be felt to be possessed in the conscious state. However it will be an illusion.

Why something rather than nothing?

TO SOME, the above question has little force. Though one might be impressed more by the existence of a black hole than by that of a stone (given the exotic nature of the former compared to the latter) it need not be necessary to be impressed by the sheer existence of something-or-other. Why should the 'natural thing' be the occurrence of nothing, while the special be the existence of anything? The philosopher Bergson expressed this qualm succinctly: 'The presupposition that *de jure* there should be nothing, so that we must explain why *de facto* there is something, is pure illusion. The idea of absolute nothingness has not one jot more meaning than that of a square circle.'

Yet there still does seem a problem to be faced. For we have looked at the other impossible questions so far in this chapter and arrived at answers which, whatever their validity, are based on the notion of cause and effect: emergence of the Universe as a quantum fluctuation from the vacuum, stretching out of time, one TOE explaining the next, consciousness arising from brain activity. Admittedly some of these explanations involve an infinite regression, but each step of the regression allows causality to be seen to be effective. It would be surprising then if, after having to go to sometimes infinite lengths to preserve causality, we drop it now when faced with what looks like the most difficult question of all: 'Why is there something rather than nothing?' Let us look further to see if causality can be saved in the face of this enormous challenge.

In order to begin to answer the question it is necessary to be clear what is meant by the terms 'nothing' and 'something'. By 'thing' one could mean a sentient being like ourselves, or it could be inanimate objects like chairs or tables. It could be taken to mean stars with attendant planets, or galaxies of different types in various stages of evolution. Yet all of these candidates for 'thinghood' are imprecisely specified. We cannot expect to give a justification for the existence of any object. The question we are asking therefore is about the existence of specific objects, such as the plates or other particular objects in a kitchen.

The problem of describing the particular objects of the Universe whose existence we wish to derive must now be faced. Only when such descriptions are given will we have specified the objects to be discussed. At what level is this description to be made? Do we use a simple photograph taken with a hand-held camera, or should we be more precise and use electron micrographs with thousandfold or more magnification? Should we use a classical description or a quantum mechanical one? Do we need to go down to the quarks which make up the object, or even further down to the supposed superstrings? If there is an infinite set of scientific theories each one demolishing the preceding one, but the chain never ending, as claimed previously, at which level of these theories should we stop, if at all? For the question is one of the level of precision at which we should describe objects and at which we should expect an explanation of their necessary existence. Such a question arises in the comedy *The Bald Prima Donna* by Eugène Ionesco, with a Mr and Mrs Smith meeting and finding that they are living in the same house, sleeping in the same bed and, they think, have the same child. However at the end of the play the maid tells us, the audience, that the couple are mistaken in their relationship, for the child of each of them has one blue eye and one brown eye, but one child has its left eye blue, the other its right eye blue. At that level of exactness any proof of the necessary identity of the child of Mr Smith with that of Mrs Smith must be wrong.

The problem we are facing is worse than that of the Smiths. As each scientific theory is displaced by the subsequent one in the chain, the preceding one is seen to be completely incorrect. It gives answers which are satisfactory for the range of phenomena for which it was previously successful, but even there those answers are only an approximation. The theory is now invalid. But so will be a subsequent theory when it, in its turn, is displaced. This implies that we cannot tolerate an inaccuracy at any level of the sequence of scientific theories because we must be able to discuss the description of objects at an arbitrarily high level of precision. Otherwise we will not be sure that they are indeed the actual objects that exist 'out there', but are not incorrectly described at a deep enough level of existence.

We seem to have painted ourselves into a corner. For infinite precision seems to be required of our description of any thing (and therefore also of 'no thing') and we can never give that due to finite time and resources. In other words we can never know the ultimate nature of things; such a description cannot even exist, since the infinite sequence of theories of everything is not expected to converge to some grand and ultimate theory but must go soldiering on for ever. That infinite regress was essential to get us an answer to the question: 'Why that TOE?' but it has now landed us in hot water over trying to answer the questions: 'Why this cup, plate, jug, etc.?' or 'Why not nothing?' Either question is about an object we cannot be absolutely clear about, since we can never describe it at all levels of precision – we would never have the time nor the energy!

Part of the pressure to give an answer to this deep question of something rather than nothing is the persistence of the scenario of existence coming out of the void. For, as Aristotle put it, 'How could there be change if there were no actually existing cause?' I would claim that this aspect makes the problem clearer. In the same way that the infinite regress of theories which I claimed earlier dispenses with the question 'Why that Toe?', so the same regress and its related sequence of ever more precise levels of description of objects dispenses with the question about existence of something rather than nothing. That is because we cannot begin scientifically to answer the latter question, but must forever burrow away at getting our description of the objects of the Universe, and the Universe itself, ever more precise. There may be features common to various levels of description, such as the vanishing total gravitational and kinetic energy of the Universe mentioned earlier, which may persist at all levels (a hint of all being nothingness perhaps?). But we cannot proceed in any precise way until the job of description is done. Nor can we appeal to a God whose hand can be discerned in the beauties of the underlying mathematical description. For we will never see that underlying TOE; it is a chimera, never to be grasped. God is not needed for mental experience, which arises from the subtlety of brain activity. He cannot be appealed to by prayer or meditation or by mystical or drug-induced experiences (as by the 'sacred mushroom' for example). Those are just going on inside the brain of the percipient and anyone who claims to have experienced God or understood the Universe in a manner impossible by the more painstaking but objective scientific approach is, I would claim, misled in making such a charge. Our wish to understand cannot be granted by leaps of faith or delusions of fantasy and hallucinations. We must use science, and that would seem to lead us to a denial of ever understanding the ultimate origin of things. We may make progress but can never arrive. If that is the nature of reality, so be it. At least we should be clear as to how far our scientifically aided thought processes can ever take us. But the limitation is really in the nature of reality, not in ourselves.

To conclude, our answer to the fourth impossible question, 'Why is there something rather than nothing?', is that on the basis of a conjectured infinite regress of theories of Matter we can never answer this question since we will never know the precise nature of the 'something' which we would like to predict around us. This really is an impossible question; in principle we will never be able to answer it.

PART TWO

THE MATERIAL UNIVERSE

Chapter Five THE NATURE OF MATTER

Delving down

THE PROGRESS made so far in understanding Matter is the great success story of modern science. The naïve reductionist approach of explaining the properties of an object in terms of the behaviour of its parts, and their properties in terms of their parts, and so on, has worked remarkably well. The resulting sequence of parts, sub-parts, etc., appears to have an increasing degree of simplicity and elegance about it. Fewer different parts are needed to explain ever greater ranges of phenomena involving material objects as we probe deeper and deeper into matter. This amazing advance has even led some scientists to claim that fundamental science will soon come to the end of the road and the 'ultimate' parts – the truly fundamental particles, dressed in their final elegance – will emerge. These particles will satisfy some laws of force which will be discovered at the same time. Altogether, the particles and their forces will constitute the elusive theory of everything or TOE, as was mentioned in the preceding chapters. This holy grail has spurred on the searchers for ultimate elementarity.

Naïve reductionism has not been just pie in the sky. It has also led to technology which has made modern life so much more comfortable – the products of electromagnetism, such as electric light, telephone, radio and television; the understanding of the nucleus giving atomic fission power and (in the near future) fusion power; flight and space flight from gravity and hydrodynamics, and many other examples. Without reductionism life would be much less comfortable.

Yet in spite of the apparent simplicity being revealed by the reductionist programme, increasingly complex behaviours are being discovered in nature. The analysis of complex dynamical systems (especially involving chaotic motion) has become an important part of modern science. I described certain aspects of this in Chapter 1. But this study does not seem to require any deviation from the reductionist programme still being vigorously pursued. It only necessitates the further parallel development of the new branch of science – complexity and complex dynamics – which is already well into its late childhood. In a sense, reductionism is for the lazy, since it avoids the complex problems left behind when plunging ever deeper into Matter to discover the simple. How far has reductionism progressed in its attempt to escape the complex and reveal the simple in nature?

Formally, the molecule is the smallest recognizable particle of a chemical compound or of an element in a free state. Most normal Matter is made of molecules, which are held together by cohesive forces to constitute the vast range of material objects we see about us. Molecules must be very small to give that variety. Water molecules, for example, are about a hundred million times smaller than a beaker of water. However, there can be much larger molecules than those of water, for instance those of DNA or of polymers.

There are an enormous number of different molecules and it seems a hopeless task to bring order and understanding to this complexity. Yet it is exactly that which was brought about by atomic theory, which goes down to the next level of size of basic objects:

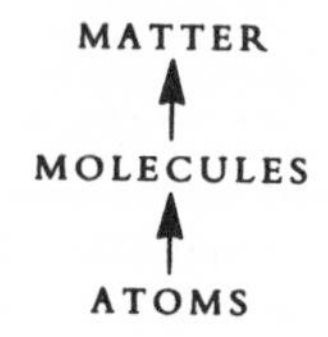

The bewildering variety of molecules can now be explained as being constructed from a much smaller set of underlying atoms.

The original idea of the atomic constitution of Matter dated back to the early Greek thinker Democritus, but it was not until the last century that it became possible to explain the enormous array of molecular combinations of atoms in terms of the Periodic Table. This arranged the different known atoms in groups according to their similar abilities to combine with atoms of other elements. The atoms were also arranged in order of their weights. The Table originally contained ninety-two different atoms of the so-called elements; a few more which do not occur naturally on earth have since been created artificially by scientists. Curiously enough, the actual existence of atoms was not accepted by some scientists until quite recently; it was not until 1960 that single atoms were photographed by means of a suitably sensitive microscope, with a magnification in millions.

Here, then, was an important watershed: all Matter was made up of only ninety-two different elements. However, these elements were not yet elementary, for they had internal structure. It was shown by Lord Rutherford in 1914 that the atom itself is composed of a central, much smaller, positively charged nucleus, in which most of the mass of the atom is concentrated, surrounded by a cloud of enough orbiting negatively charged particles (called electrons) to make the total atom electrically neutral. We have, then, the next level down:

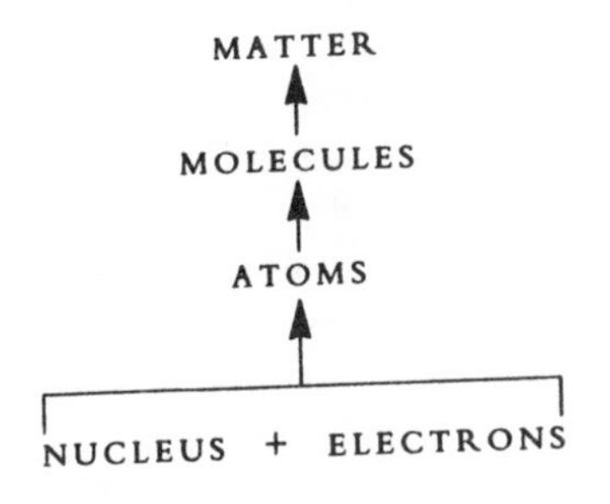

The electron had been discovered at the turn of the century as a further constituent of all Matter and its fundamental place in the atom was clarified by Rutherford's results. This led to a beautiful quantitative description of the so-called 'nuclear atom' which helped explain some of the details of the radiations emitted by atoms when they were heated up. This was based on the idea of electrons only being able to live in a stable manner in a discrete set of states (with particular velocities) as they orbited round the nucleus. Concepts of such 'special states' were *ad hoc* and contradictory to the laws of Matter known at that time, but they worked. The need for such states in fact contained in itself the seeds of an amazing revolution which was soon to sweep away the well-established laws of classical physics originally put forward by Sir Isaac Newton over three hundred years before.

Uncertain answers

THE REVOLUTION which swept away the classical physics of Newton is still with us today, and the interpretation of the new framework for describing Matter – that of quantum physics – is still being hotly debated. Since it is at the root of recent attempts to understand the creation of the Universe, and also since it is essential in appreciating the recent, supposedly rapid, progress towards the discovery of the TOE, we must spend a little time in exploring it.

The basic paradigm changes in going from Newton's physics to quantum physics are essentially two. To Newton, Matter consisted of billiard balls moving around in a completely certain manner, following their straight line trajectories unless they collided with other billiard balls. To quantum physics, as developed in the mid-1920s, the billiard balls had only a certain probability of moving along their classical paths. The heavier the object the more certain they were to follow a single path, but for particles as light as electrons their motion could spread out like a wave rippling over the surface of a pond when a stone is dropped into it. That is why quantum mechanics is sometimes termed 'wave mechanics'. Thus in quantum physics Matter behaves like a wave, the wave itself describing the probability of finding the particles. The Matter wave could produce an interference pattern if it went through a very narrow opening (as also occurs for light waves) and such effects were observed in the 1920s and have been verified many times since then. The uncertain nature of Matter can be psychologically disturbing. Albert Einstein, for example, did not like it and said, 'God does not play dice with the Universe.' It seems so contrary to what we observe that it is still being hotly debated.

I have described the first paradigm change, with certain billiard balls turning into Matter waves of uncertainty. The second paradigm shift in going from classical to quantum mechanics is as startling. In the classical world billiard balls move around smoothly. All paths are continuous. There are no 'jumps', where a particle has to be at one point and then suddenly at another. Yet in quantum physics discreteness and 'quantum jumps' are crucial elements. Energy exists in packets or 'quanta'. I mentioned at the end of the last section that electrons in the nuclear atom were supposed only to move around the nucleus in a discrete set of states. The new quantum mechanics gave support to this, but led automatically to the problem of quantum jumps. These arise when an electron in one state jumps to another. If the new state of the electron has less energy than it had before, the amount it loses is apparently emitted as light or similar radiation. But quantum mechanics could not describe what happened to the electron as it sped from one of its special states to the other.

In fact there need be no real problem if one keeps strictly to quantum ideas and does not think in the pre-quantum mode. This seems to be true of the numerous 'paradoxes' which were discovered in the years after the advent of quantum mechanics. Thus the movement stage in a quantum 'jump' as such would not exist; the electron of the previous paragraph would only have existed in its initial or final state, but not 'in between'. These difficulties may be seen as arising mainly because it is difficult for us to clear from our minds the strong impressions and prejudices of the way the world is composed in the large, and as so accurately described as being composed of sets of billiard balls moving with 100 per cent certainty along their paths. Indeed this happened to Albert Einstein, the discoverer of the discretization of energy in 1905 in the so-called photo-electric effect. In this, the detailed character of the emission of electrons from a metal surface when light is shone on it can only be explained in terms of the postulate that the light energy absorbed by the metal is carried in discrete packets called 'photons'. In spite of Einstein having been one of the early creators of quantum mechanics he never accepted its probabilistic description of an individual system such as a single electron, and had a long series of debates in the 1930s about this feature and the paradoxes it produced. His opposition to the probability aspect of quantum physics left him isolated in his later years and his search for a unified field theory for all Matter and gravity can now be seen to have failed due to his lack of account of the successes of the burgeoning quantum approach.

As might be expected, various theories have been proposed which are closer to our certain 'billiard-ball' view of the world. In 1982 experiments on light coming from the decay of a single atom were used to show that the strongest candidate as a replacement for quantum mechanics was excluded. In spite of such efforts, some still going on, there presently seems no alternative to going down the quantum path. This is particularly so because that path has been followed very successfully for a long distance, as I will indicate shortly.

Besides the unease at the element of uncertainty gnawing away at a deeper level at our apparently safe and certain classical world, there are specific paradoxes which have led to much debate over the last few decades. One of these is associated with the names of Einstein and two of his colleagues, Podolfsky and Rosen, and is succinctly termed 'the EPR paradox'. It involves the apparent ability to communicate at a speed faster than light. Since this is something which disagrees with sensitive experiments (as we will describe in the next section) it would appear to be problematic. A similarly bizarre phenomenon arises with the so-called 'Schrödinger's cat' paradox. In this, a (hypothetical) cat is placed in a chamber together with a bottle of cyanide, a radioactive atom, and a device which will break the bottle when the atom decays. Every time we look into the chamber we will see either a live cat or a dead one. However after the time at which the atom has about a 50 per cent chance of decaying, the principles of quantum mechanics lead to the conclusion that the system is a strange mixture in which the cat is half dead, half alive. Which half is which is not clear, nor how the cat would feel about this!

These and other paradoxes arise from the sometimes unconscious requirement that the rules of quantum mechanics give a description of the motion of individual particles, say particular electrons or photons (the particle of light mentioned earlier). But that is not required by the experimental successes of the theory. More correctly one is required to regard wave mechanics as describing the properties of an ensemble of identically prepared systems. This is termed the 'ensemble interpretation of quantum mechanics'. The cat paradox is resolved by such an approach since indeed we expected, as was noted, only to see dead or alive cats 50 per cent of the time (if repeated many times, each with a different cat). It is not a single cat which is half alive, half dead. This result agrees with what the ensemble interpretation of quantum mechanics would predict. The other paradoxes can be handled in a similar manner, although some of the resulting interpretations cause us to realize the bizarre character of such phenomena, even if no true paradoxes remain.

No faster than light

ANOTHER OF THE important developments of our understanding of Matter in the first half of the present century was the sole creation of Albert Einstein. He called his theory 'Special Relativity', and it involves the way in which different people, possibly moving with respect to each other, can describe the differences between each other's observations which they make on the same system. The word 'special' indicates constraints on the way the different experimenters can move – only at constant speeds relative to each other. The term 'relativity' denotes that different observers are comparing their experimental results with those of their colleagues and all is relative. No observer is singled out over all the others.

As with quantum mechanics, special relativity entailed a paradigm shift. This involves how speeds of moving objects are to be combined. If you throw a ball forward from a moving train you will expect the ball to travel at a speed equal to the sum of the velocity of the train and the speed at which you throw it from the train. That is correct for the slow speeds involved. But when an object is travelling close to the speed of light, then if it ejects another object with a forward velocity also close to the speed of light we might expect the total speed of the ejected object to be nearly double the speed of light. Not so, claimed Einstein: no object can travel faster than light. This response was based on the assumption that light always travels at the same speed, no matter how it is produced or observed. Such a claim has been validated experimentally over an enormous range of wavelengths of light and its mode of production.

Let us investigate the effect of the requirement of the constancy of the speed of light on the measurement of time. That can be done, for example, by a light clock. This is constructed from two mirrors opposite each other. A pulse of light bounces from one to the other and back again, each return trip being a unit of time of the clock, which we will call the tick tock. At every tick and tock, light bounces back and forth. Suppose the clock is now moved sideways. Looking from above, a stationary observer would see a pulse following a zig-zag path. This new path will be longer than that followed when the clock was still. Remember now Einstein's condition that light always travels at the same speed. The implication of this is that it takes longer for the pulse to bounce back and forth when the clock is seen to be moving than when it is still. A stationary observer looking at the moving clock will see it making time as tiiiiick-toooock, while an observer moving with the clock will observe the usual tick-tock. In other words the moving clock will be observed (by a stationary observer) to be going slow compared to any of his stationary ones. Hence Einstein's dictum 'Moving clocks go slow'.

As for quantum mechanics, the new paradigm leads to unexpected results. A well-known one is the twin paradox, in which a twin ages less if he travels at high speed to a distant star and then returns to earth with similar rapidity, compared to his stay-at-home twin, who never left the earth. There are numerous other bizarre consequences of special relativity, all of which have been verified with great accuracy. In particular, it is accepted that the speed of light is a barrier which cannot be broached; one cannot travel faster than light. This has serious implications for space travel since it limits how fast we can ever travel to nearby or more distant stars, but we do not know of any violation of this result to give a loophole. Another famous result of special relativity is the identification of mass and energy '$E = mc^2$', which ushered in the nuclear age we are still living through. An important consequence for our present odyssey into Matter is that special relativity becomes more important the smaller and lighter the particles discovered in the quest. That is because the lighter particles can be accelerated more easily to high speeds. In spite of accelerating electrons ever closer to the speed of light, no breaching of the light barrier has ever been discovered. This is quite fortunate since if one could travel faster than light then the following limerick could become true:

> There was a young lady called Bright
> Who could travel much faster than light.
> She went out one day,
> In an Einsteinian way,
> And returned on the previous night.

Such problems have preoccupied many science fiction writers, and are the theme of such entertaining films as *Back to the Future*, in which the hero causes his siblings to begin to disappear by his actions in their past. It would indeed be disturbing for science if it were physically possible to go back and interfere with one's past (such as preventing one's mother and father from ever meeting!)

In spite of careful searches for tachyons (*tachys* is Greek for swift), the hypothesized faster-than-light particles, no trace of them has been seen. Nor has any violation of causality (cause always precedes effect) ever been observed right down to the shortest distance probed inside the elementary particles.

Probing deeper

THE PHENOMENON of radioactivity has been known since the careful work of Madame Curie and her colleagues at the end of the last century. Various types of radiation (termed alpha, beta and gamma rays) were discovered to be emitted from heavy atoms such as uranium, and the remaining material changed into lighter elements such as lead. Some of the radiation involves beams of nuclei of helium (which must come from the atomic nuclei) and one can only conclude that the atomic nucleus is not inviolable but can break up if too big. It was also found that nuclei of atoms were almost exactly a whole number of times more massive than the proton. For example, the helium nucleus is about four times as heavy as a proton. Yet it only has twice the proton's electric charge, so it could not be made up of four protons. It could, however, be composed of two protons and two electrically uncharged particles as heavy as the proton. Such new particles could also be used in all the other nuclei to explain their masses and charges. This led to the model of the nucleus of any atom as being composed of a set of protons and of almost identical, but electrically uncharged, particles called neutrons (discovered in 1932). The phenomenon of radioactivity was also explained in terms of the emission of either helium nuclei (alpha rays), electrons and a neutral, massless companion called the neutrino (beta rays), or light (gamma rays). The neutrino's existence was first postulated in the 1930s in order to explain the detailed energies of beta-ray emission, and it was observed experimentally several decades later. This gives us the new reductionist picture:

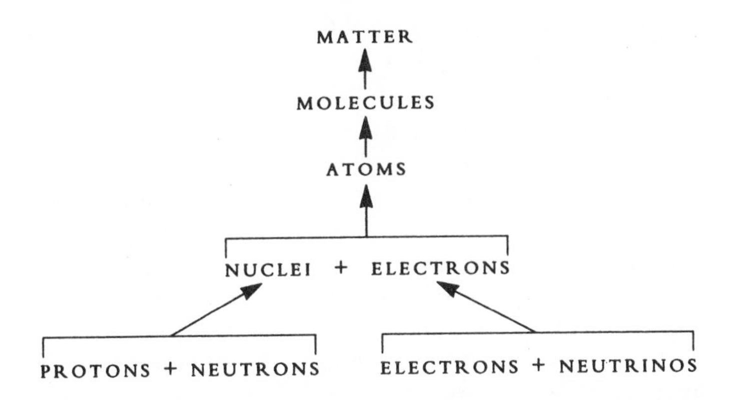

At that point some scientists would have liked to stop. There were the four elementary particles (proton, neutron, electron and neutrino) which together with the quantum of light, the photon, were a manageable set hopefully explaining all the phenomena of Matter. But that was not to be. The weaponry needs of the Second World War had produced a new technology, and with highly sensitive microwave devices and strong magnets a new breed of particle accelerators was built after the war, probing Matter to increasingly higher energies and shorter distances. The new machines accelerated beams of protons very close to the speed of light and slapped them on to targets of hydrogen or similar materials. In this way a host of completely new particles was discovered, either as heavier but short-lived companions to the proton and neutron, or as heavier (but also short-lived companions) to the photon. These new particles were not completely unexpected. In the mid-1930s a Japanese physicist had argued most creatively by analogy that if photons are the 'glue' of the electric force holding electrons to the nucleus then there must be a similar glue holding the neutrons and protons together inside the nucleus. This 'glue' he called 'the meson', and indeed some of the new particles exactly fitted the job description. But others were very different.

To understand the difference between the good guys and the bad guys we must take note of the fact that the particles I have introduced so far all have an intrinsic quality – their spin. They are spinning tops. Some of the top-like particles can either spin upright or upside down, so they have two possible spin states. Others have one spin state only, and yet again others have three spin orientations (up, down, and sideways). The photon has been found to have three spin states and it was realized that the glue to hold the nucleus together must be carried by particles, like the photon, with one or possibly three spin states. But some of the newly discovered particles had only two observable spin states. They were closer in properties to the electron or neutrino, with two spin states, so could only act as Matter itself – not glue.

After many painstaking and difficult observations it was found that there were two sorts of new particles being produced in the particle accelerators. Firstly there were two new companions to the electron and its neutrino, which are now called the mu-meson (and the mu-neutrino) and the tau-meson (and the tau-neutrino). Then there were considerably heavier particles which could be regarded as companions of the proton and neutron, some of them ultimately decaying into the proton. Those companions which all had the same mass could be assembled into groups of octets (eights) or decuplets (tens). Where there were gaps in an octet or a decuplet, predictions were made as to the nature of the missing particle. Such candidates were later discovered with the requisite properties in the debris spewed out of the collisions between the proton beams and their targets in particle accelerators.

This preponderance of eights and tens led to the approach being termed the eightfold way and caused some physicists to disappear into Eastern mysticism (which also favoured an 'Eight-fold Way' to salvation). They reappeared later saying it had all been known by the Eastern sages they now worshipped. Sanity was preserved, especially with the quark hypothesis in 1964, which proposed that there is a triplet of quarks out of which this plethora of new particles could be constructed, as if by a lego construction kit. This kit used the magic formulae

$$3 \times 3 = 1 + 8$$
$$3 \times 3 \times 3 = 1 + 10 + 8 + 8$$

to explain the eightfold way, and how the proton and its companion are made up of quarks. It seemed, after all, that three was a lucky number for nature as well as for my superstitious friends!

These formulae are taken to read that two quarks occurring together in some particles (the 3 × 3 combination on the left-hand side of the first equation above) can make up an octet of particles (the eight on the right-hand side of that equation) and three quarks (the 3 × 3 × 3 combination in the second formula) in a proton-like particle, say, can form eights or tens (as found on the right-hand side). That was in excellent agreement with the observed grouping of the new particles. It was also possible to explain from this model many detailed features of the way particles should scatter from each other if they are made up of quarks. This success has led to universal scientific acceptance of the quark model of the particles. The plethora of new particles had thus been explained in terms of a very few constituents. Finally, the glue to hold the quarks together inside particles had to exist. This was termed 'gluons', of which eight separate sorts were predicted. Evidence for these has also been obtained over the last decade.

So far I have tried to present as simple a picture as possible of the 'elementary particles'. However, a little more description is needed to flesh it out. Thus to fit the charges of observed particles there had to be at least two different triplets of quarks, the up-quark u, with electric charge 2/3, and the down quark d, with charge 1/3. The unit-charged proton and electrically neutral neutron then results from the combinations:

proton = (u u d)
neutron = (u d d)

The force between these quarks – the gluons – would then be expected to cause these combinations to be formed. To explain all of the new companions to the proton in eights and tens mentioned earlier, four more quark triplets – charmed (c), strange (s), bottom (b) and top (t) – had to be proposed. Only 't' has so far resisted observation, most likely due to its being too heavy to be produced by present accelerators. It is now possible to fit all of these new particles together into three families composed of:

$$(u, d, e, \nu_e), (c, s, \mu, \nu_\mu), (t, b, \tau, \nu_\tau)$$

which, together with the photon (γ) and eight gluons (gl), is thought to explain nearly all the observed particles.

Finally, then, we extend the reductionist 'ladder of descent' into the land of the very small as:

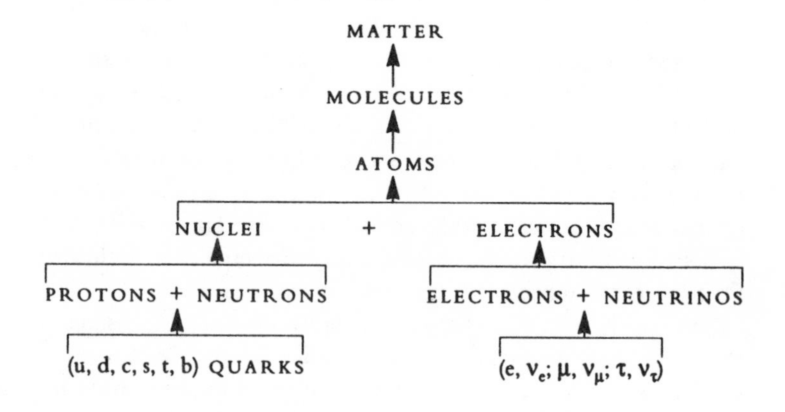

Gluing it all together

WHILE WE HAVE concentrated on the particles we have only occasionally mentioned the forces between them, as if by chance. These forces themselves must be explained and also provide very important further clues as to the nature of Matter.

Imagine yourself as a Martian circling the planet in your UFO. As you speed over Wimbledon during the two weeks of tennis, you stop above one of the courts (high enough to avoid disturbing the play and the crowds) and observe carefully. You note the passage back and forth of tiny white (or yellow) spherical objects. While that proceeds, many robot-like objects stay in their places, except for the turning backwards and forwards of their topmost portion, as if visually to follow the passage of the little white spheres (in the process you make a mental note that the machine vision system used by the robots must be very poor to require such excessive motion and energy use in trajectory tracking). When there is cessation of the back and forth movement of the white objects the robots themselves move, possibly to leave the region altogether. You find this a most interesting phenomenon, and on returning to Mars report back to your superior that there must be an attractive force caused by the exchange of the little white objects, which ceases soon after the exchange stops.

Particle physicists envisage the forces of nature in a similar manner as being brought about by the exchange of particles or quanta. Thus the electromagnetic force between charged particles is due to the exchange of photons between them. Similarly, quarks are bound together inside the proton or neutron by the particles of the nuclear force field, and these are called gluons. There is one other force we have mentioned, that of radioactivity. Its nature was probed in the 1930s but it was not properly understood in the above language of quanta being exchanged until the unification of electromagnetism and radioactivity was attempted in the late 1960s. It was suggested that quanta had to exist for the radioactive force, which could be regarded as 'radioactive' companions to the photon. These were mesons (with three possible spin states, like the photon) called the W and Z mesons. Their masses were predicted to be about ninety times heavier than the proton. The new LEP (large electron positron machine) was specially built in Geneva to detect the W and Z, and indeed they were found with properties exactly as predicted. Now there are tens of thousands of Zs, as well as many Ws. The importance of the former is that they can tell us, as can be shown by fitting properties of the Z to various models of the other particles, the maximum number of families which contain light neutrinos; the answer is three.

Our story about Matter is almost complete. The reductionist ladder is now seen to be:

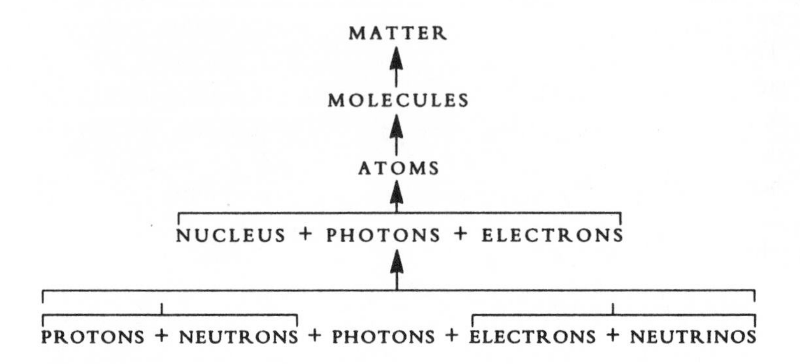

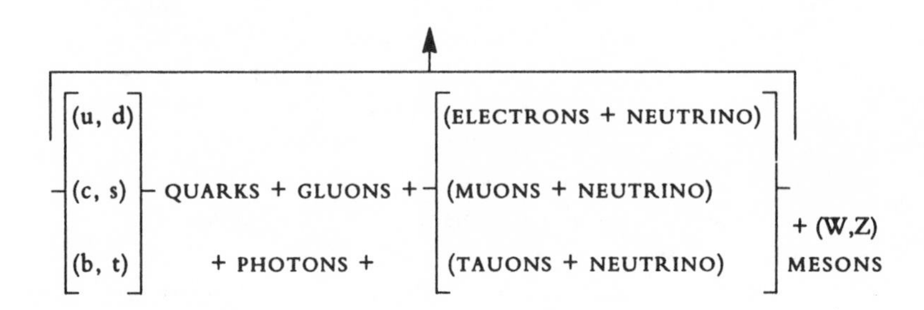

The quanta of the forces of nature are seen to be:

Force	Quanta
Electromagnetism	Photon
Nuclear Force	Gluon
Radioactivity	(W,Z) mesons

The reductionist ladder and the table of quanta for the forces are the latest available at the time of writing this book. Perhaps by the end of the millennium new results from particle accelerators presently being built will add further to the ladder. New entries may be discovered for the force table. But they can only increase the success of the reductionist programme; there is no chance that the ladder can be changed higher up. It can only be extended at the bottom.

There is still one important aspect to describe. It is the thrust towards unification, which runs as a thread through all of the work described so far. It began as far back as 1864 when James Clark Maxwell unified electricity and magnetism. These were further combined with the chemical forces by the nuclear atom in about 1910 (with understanding of the Periodic Table gleaned thereby). Further unification with radioactivity was achieved by the W and Z mesons. This led to the so-called electro-weak force (the epithet weak because radioactivity is the weakest of the three non-gravitational forces). Possible unification with the nuclear force, now called chromodynamics (by analogy with electrodynamics) has been attempted, but has not yet led to any successful model. The story of unification so far can be presented thus:

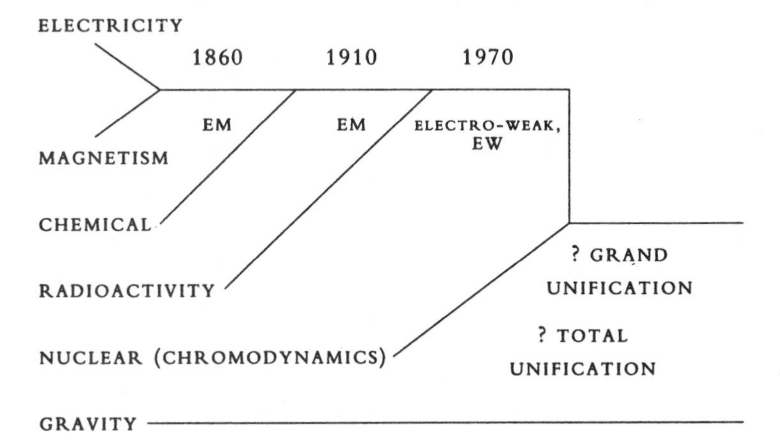

An important concept which helped to guide the advance of our understanding of the forces of nature is called the 'principle of gauge invariance'. This allows an enormous reduction in the possible forces that need to be analysed. Let us return to the picture of the Martian over Wimbledon and the tennis balls being exchanged between robots. In addition to what is visible externally, a very clever Martian would include, according to the gauge invariance idea, circles on the heads of the playing robots with a moving dial hand on each. The direction of this hand is seen to change suddenly as a tennis ball bounces off the robot. The dial-hand directions appear to be involved with the force between the robots. Yet more careful analysis by the Martian shows that the force between the robots caused by the exchange of the tennis balls is independent of the dial position. The independence of the force on the dial-hand direction is a close analogy to the gauge 'invariance' principle. Careful analysis shows that this principle is extremely powerful in restricting the form that forces can take; the electro-weak and chromodynamic theories were constructed mainly using this as a guide. Possible grand unified theories (unifying these two) were also suggested in the gauge-invariance framework, and are still being tested.

We have now added gravity as the fourth force of nature in the unification picture and drawn in its putative combination with the other two. That was already indicated by a question mark in the unification diagram. If we could achieve such total unification of all the forces of nature then might we not finally have discovered the TOE?

Chapter Six

MORE WEIGHTY MATTERS

Gravity as geometry

GRAVITY IS THE force which man has undoubtedly experienced consciously longer than any other but yet has had most difficulty in understanding and modelling. Primitive man had always had gravity to contend with in avoiding falling over cliffs or in using steep hills to give him advantage over his enemies. Catching a bird or an animal needed understanding of the effect of gravity on the missile used to bring down the prey. Early builders had to learn to combine solid shapes together carefully so as to build habitations able to withstand this ubiquitous force.

The search for an understanding of gravity had an auspicious beginning. At the same time that Newton discovered his three laws of motion he devised a universal law of gravitational attraction between bodies called 'the inverse square law' (so that if two bodies double their distance apart their gravitational attraction reduces by a quarter). Together, the laws of motion and the inverse square gravitational force led to a well-defined model for the motion of the planets round the sun; this was found to give a remarkably accurate set of orbits for them and even allowed the discrepancy of Uranus in its predicted orbit to indicate the existence of a further perturbing planet. This was searched for and found; it is Pluto, the ninth planet.

The inverse square law of gravity began to appear to contradict other physical principles at the turn of the century. One very general concept was that of a limit to the speed of propagation of any effect. Albert Einstein had enshrined this in his principle of special relativity, which we described in the previous chapter. The basic premise was that light always travelled at the same speed, and implied that the velocity of light was the ultimate speed of all objects. However, gravity seemed to be an instantaneous force between two massive objects, so contradicting special relativity.

Einstein realized that he had to extend special relativity to the case when two observers are not just moving at constant speed with respect to each other (as in special relativity) but are even accelerating. He considered the case of someone in a lift in a tall building. If the lift is stationary the person experiences gravity like everyone else. If the lift cable snaps and the lift goes into free fall then the person becomes weightless. The force of gravity on him has been effectively removed as he is accelerating to the bottom of the lift shaft. The phenomenon of true weightlessness has now been experienced by many astronauts, but in the early 1900s this was purely a 'thought experiment', and remarkably creative. For it led Einstein to realize that gravity was equivalent to acceleration. This meant that, at least in a small region, the effect of gravity could be nullified by a suitable acceleration – that of 'free fall'. However the converse would be true: if one could extend special relativity to include observers accelerating with respect to each other then one or other of the observers would be expected to be experiencing what they would interpret as a gravitational force (since gravity and acceleration were seen to be equivalent).

One of the most important conceptual developments of special relativity was the fact that, like space, time is not absolute but depends on the observer. The similar character of these features allowed for the introduction of the so-called space–time continuum, with three independent directions of space and one of time. This continuum was populated, in Einstein's special relativity, with observers moving around with constant velocity. None of them had a privileged character, all was relative, hence the epithet 'relativity'; 'special' denoted the restriction of the observers to constant speed. With Einstein's new understanding of gravity as acceleration came the realization that one had to consider observers moving in an arbitrary manner. The notion of relativity was still important but now was extended to cover arbitrarily moving observers.

At this point Einstein introduced an approach which changed the face of physics and of the way we look at the world. He proposed that the space–time labels we attach to events are irrelevant to the underlying processes going on. They could be looked at by observers moving arbitrarily, who should still give the same account of the underlying forces at work. Thus the person either standing in the stationary lift or undergoing free fall would not be able to detect any change in the underlying forces of nature if they could not look out of a window. The only true effect of gravity on the surface of the earth would be observed by a slight difference in the precise direction needed for free fall at one side of the lift compared to the other because of the earth's curvature.

'Labels are unimportant' was the theme of the new General Theory of Relativity. If that were so then how could gravity produce its effects? Einstein realized that the only way to tackle this question was in terms of the geometry of space–time. Geometrical properties are those that do not depend on labels or co-ordinates describing where objects are in a space. In geometry the concepts are points, lines, planes, surfaces and so on, and the properties investigated are, for example, whether three lines intersect in a point. None of these properties depends on particular labels attached to the points, lines, etc.

There is no reason for the four-dimensional space–time (of three space and one time directions) to have the properties originally suggested by the Greek mathematician Euclid. In Euclidean space, often termed 'flat' space, two parallel straight lines never meet. Also, the sum of the interior angles of a triangle is always 180 degrees. However there are many spaces which are not flat. Consider, as a simple example, the surface of a sphere. An analogue of a straight line (which describes the shortest distance between two points) is a great circle on a sphere. Any two such great circles always meet, whereas two parallel lines in the same plane never meet. Another difference is that the sum of the interior angles of a triangle on the sphere is always greater than 180 degrees and could even be 360 degrees. It had been suggested by various mathematicians in the last century that our physical space may not be flat and one had even tried to find out by a survey what was the sum of the interior angles of a triangle formed by three mountain peaks. That failed to show that physical space was not flat, and it was not until Einstein's creative leap that physical curved space became a reality. He conjectured that our four-dimensional space–time is curved and that the curvature itself (defined in a correct geometrical manner) was the gravitational force acting on light bodies moving freely around in the space–time.

One of the reasons why such a simple description of the way small bodies moved under gravity was possible was the 'free-fall effect' mentioned earlier. Weightlessness implies that a person's mass is unimportant in determining how they move under gravity. This feature is called 'the principle of equivalence', and may be expressed as 'all particles fall with equal acceleration under gravity'. It seems to hold for all Matter, even down to protons, neutrons and electrons and has been tested to an accuracy of one part in one hundred thousand million.

The idea of gravity as curvature of space–time can be understood in terms of 'rubber' snooker. Here the snooker table top is made of rubber. A sneaky way of trying to win is to have the coloured snooker balls made of lead. These will stretch the table top near these balls, so leading to curvature of the surface. Your unsuspecting opponent will play a shot which will be distorted as his own lighter ball curves towards the heavier coloured ones he may want to miss. Of course if he also is sneaky enough to play with a lead ball then it may not be so easy to predict how much it will be affected by the distortion of the surface caused by the heavy coloured balls. This corresponds to the situation in which the original curved space–time is itself further distorted by the test particles moving through it.

To be able to work out how much curvature of space–time is actually produced by a massive body, Einstein introduced in 1915 his 'equations of General Relativity'. In this he equated the geometrical curvature of the space–time (effectively the force exerted by gravity) to the distribution of energy and its flow throughout the space–time produced by the Matter. In other words the equations of General Relativity are of the form:

Curvature of Geometry = Energy of the Matter Distribution

For a very massive body like the sun it is possible to solve the Einstein equations and obtain a special solution which gives rise to Newtonian effects at a great distance. However, on closer inspection there are differences in the results of the Einsteinian and Newtonian approaches. In particular, photons of light have energy and therefore Mass (from '$E = mc^2$'), and so should be deflected by moving near the sun. In 1919 a total eclipse of the sun was carefully studied by a team of scientists, in particular looking at a star whose light passed very close to the sun at eclipse. By knowing the star's correct position and its apparent position (observed at the sun's edge when the sun was almost fully eclipsed) it was possible to calculate how much a ray of light from the star had been pulled towards the sun as it grazed the sun's edge. The value was in almost exact agreement with Einstein's predicted value, and even better agreement has since been obtained. Multiple images of very distant quasars (distant compact galaxies emitting enormous amounts of radiant energy), due to nearer galaxies in the line of sight acting as gravitational lenses, have been reported. Further tests of Einstein's General Theory of Relativity have been carried out since then, such as that of the predicted amount of slow rotation of the orbit of Mercury around the sun, and the variation of the delay of radar signals reflected from planets, as they travel round the solar system. In all of these tests General Relativity has been found correct to the limits of experimental accuracy. The level of agreement is now regarded as certain to a tenth of a per cent. Newton's inverse square law has been displaced; the Universe is not a flat Euclidean three-space (absolute space) with an absolute time running along smoothly and independently, but is a curved space–time continuum in which the space and time are inextricably intermixed and relative, and gravity is the curvature of the geometry determined by the Matter distributed throughout it.

Black holes

IF YOU WANTED to buy a car you would first test it out by taking it for a short run on the road. The more severe the trial run, the better your faith in the car. If you want to accept a scientific theory you should first test that out by taking it for crucial trial runs. In each trial the prediction of the theory for a particular phenomenon is first evaluated and then compared with the actual results obtained by suitable experiments. Again, the more severe the experimental conditions the more faith one will have in the theory if it has made the right predictions. 'Gravity is the curvature of space–time.' This was the conclusion of Einstein's thesis in 1915 and it has captured the imagination of generations of people ever since. It is an elegant solution to the problem of the nature of gravity and leads on to enormously important cosmological implications, which we will explore in the next chapter. But it also produces a very disturbing possibility even on a small scale: that of the creation of what are called 'black holes' and the crushing and subsequent disappearance of Matter to a point at their centre. This leads to a paradox which has ultimately to be resolved – how can one avoid the disappearance of Matter in the face of strong gravitational fields? It is these conditions which appear to give the General Theory of Relativity its severest test. Since the predictions immediately disagree with the model of Matter used in the Einstein theory in the first place (in which Matter cannot just disappear), then Einstein's model seems to have failed its test.

Let us look at this in more detail to see exactly what is at stake. The existence of black holes is not a new possibility. They were first predicted at the end of the eighteenth century, using the 'corpuscular theory' of light to suggest that a large and concentrated enough mass would prevent any light from escaping from its surface and any light falling on it would always be trapped. But this was only derived from a variety of different ideas. The further properties of black holes, however, were not discussed until some time later. Such a fantasy became more of a reality with Einstein's ideas on gravity and light and also with the increasing understanding of the evolution of stars. In particular came the appreciation that the source of energy in stars comes from nuclear fusion and that there is a decreasing chance of this occurring as a star uses up all its nuclear fuel. Initially, hydrogen atoms are fused to helium and energy is produced; then helium atoms fuse to produce carbon and the concomitant energy. This process of fusion continues until iron is formed. However, fusion of iron gives out no further energy. Ultimately then, any star must die when it has changed a large enough proportion of its original hydrogen material into iron. Its ending could be quite catastrophic, considering that the star's volume is caused by the enormous pressure exerted by the starlight trying to get out of the interior of the star to its surface. Once that pressure is cut off, when energy ceases to be produced by fusion processes in its interior, the star would collapse on itself quite catastrophically, in the process blowing off a portion of its exterior. These have been observed by mankind over the millennia as supernovae explosions in the heavens. For example the appearance of the Crab Nebula in 1054 was noted by Chinese at the time.

The subsequent fate of a dying star is relatively benign if the star can lose enough of its mass so that the self-gravitational force between its parts is not too strong. It has been calculated that if it can settle down to a mass no greater than about one and a quarter times the mass of our sun, the star will collapse to be about a hundred million times as dense as before. It will slowly die out to live perpetually as a white, then invisible black, dwarf (so called because it is only about the size of the earth). It is retained at this size, in spite of no light energy trying to escape from its interior, by the forces of resistance between the internal electrons which prevent them from being pushed too closely on top of each other. If the mass of the collapsed star is up to about three times that of the sun the electrons cannot hold the remaining Matter against further collapse. The star will collapse even more, to be a million times more dense than the black dwarf, being now supported by the neutrons of which it is made. Appropriately enough, it is now called a neutron star, and is only a few kilometres in diameter. This will initially possess a strong magnetic field and, while the neutron star rotates, it will throw out oscillating radiation and behave exactly like the pulsars first observed at Cambridge in 1973.

If all this mass loss cannot be achieved (because the star was initially too heavy) then there is no way the Matter in the star's interior can support itself against ultimate collapse. Catastrophe ensues! As the material of the star collapses ever more the gravitational force around it will increase. This will cause the escape velocity from the surface (the velocity of any object to escape into orbit or beyond) to increase likewise. Thus on our earth the escape velocity from the surface is about 7 miles per second, on a white dwarf about 1000 m.p.s. and on a neutron star 10,000 m.p.s. The imploding star will ultimately reach a size at which the escape velocity is that of light. It was noted in the previous chapter that one cannot travel faster than light, so from that critical size the collapsed star would become black. No light falling in or emitted from it would ever be able to escape from the inward pull. A black hole would have formed.

The spherical surface formed at this critical size is called the 'event horizon'. This is aptly named since events happening inside it can never be communicated to the outside Universe. The existence of such a barrier to communication presents an important question to the philosophy of science. For how can any predictions made about the interior be communicated outside and checked? I was, for example, asked to adjudicate a competition to find the best essay on 'How to fall into a black hole and survive'. My difficulty was to know how to say whether the proposal was right or wrong. In principle one will never know the answer. That is, unless one falls into the event horizon oneself. But then how can one follow the usual channels of science, which involve publication in a learned journal and subsequent checking and criticism by one's peers. Unless, again, those peers fell in, with the scientific publishers, and Uncle Tom Cobbly and all! Only if all the relevant and interested scientists partook mass suicide with you and fell into the event horizon would the answers you obtained be properly validated. However, that validation itself would not last long.

As one fell into a black hole one would be stretched lengthways but squeezed ever more narrowly. If the black hole were large (say as massive as our own galaxy which is composed of a hundred thousand million suns) one could calculate that one would not notice the squeeze and stretch effect until a week after falling through the event horizon (with another week to go before reaching the centre). A black hole as heavy as our own sun would have a critical size of only 3 km, one as heavy as our earth would only be about 1 cm across. For either, the gravitational effects would become severe long before the event horizon was reached.

Having fallen through this critical surface one would be sucked inexorably to the centre, where the gravitational force would become infinitely strong. Moreover it would only take a finite amount of time to reach that central region. We must remember that time is relative, and the relevant time being used here is that experienced by the falling body – its so-called 'proper time'. A watch strapped to one's wrist would register exactly the proper time involved. For the galaxy-weighted black hole, it would take about two weeks to fall from the horizon to the centre. But when the Matter arrived at the centre it would be forced to disappear. This is the internal contradiction we mentioned earlier, since Matter cannot just disappear, at least in its form known to us on Earth. It may get transformed from one sort of Matter to another, but it does not disappear. One could conjecture that Matter behaves very differently out in the distant reaches of the Universe, but since we have no clue about that it is not a very useful suggestion.

Since there are good candidates for black holes both in our own galaxy and elsewhere, one cannot avoid the above paradox of Matter disappearing at the centre of a black hole by claiming that they do not exist. In our galaxy is Cygnus X-1, which is a binary system composed of a visible blue supergiant of at least ten solar masses together with an invisible companion (observed by its effect on the motion of the visible star) moving around it every five days, the companion being at least ten times as heavy as the sun. There is a highly variable X-ray emission, which is thought to arise from material being dragged from the blue supergiant on to the invisible partner. This latter is difficult to conceive of other than as a black hole. Another candidate is the Seyfert galaxy NGC4151, about fifty million light years away. The ultra-violet emission from it indicates clouds of gas streaming away from a compact central region which is at most thought to be thirteen light days across (a light day is the distance travelled by light in one day, about twenty billion miles). The estimated central mass of about fifty thousand solar masses makes it very hard to avoid concluding that the central region is a black hole. Finally quasars, the very distant point-like objects which are emitting an enormous amount of radiation, can only be explained as energy sources if that energy arises from the continued cannibalization of surrounding stars by a central black hole.

Even if all of the above are not, ultimately, found to be black holes, the 'thought experiment' of falling into a black hole and calculating how soon one entirely disappears is still enough to indicate that general relativity and classical matter contain the seeds of their own destruction. Moreover we are living in an expanding Universe, which seems suddenly to have expanded from nothing in the Big Bang about fifteen billion years ago. Play the cosmic film backwards and our whole Universe disappears. One cannot avoid this by claiming that disappearing Matter in very strong gravitational fields does not exist. Our Universe is the reverse proof of this. Thus the General Theory of Relativity, constructed as

Curvature of Geometry = Energy of Matter

must somehow be wrong, since it predicts the absurdity of disappearing Matter. We will explore possibilities of this in the next section.

Uncertain gravity?

THE CONCLUSION in the previous section that Matter disappears at the centre of a black hole in a finite amount of time destroys the beautiful picture built by Einstein in 1915 in his General Theory of Relativity. After the developments in our understanding of Matter described in the previous chapter it is clear that an updating of the 1915 model is needed. We must at least put in the quantum mechanical features of Matter discussed in Chapter 5. Matter might then become smeared out by the uncertainty of it being at any particular place at a certain time and so produce new effects.

As far as phenomena outside, and especially close to, the event horizon are concerned the most interesting phenomenon was black-hole radiation. This was predicted by Stephen Hawking in 1974, and corresponded to the black-hole event horizon being considered as being hot. Its temperature increases as its mass gets smaller because the effective gravitational field at the surface is fiercer the smaller the black hole. Black-hole radiation should therefore be strongest from mini black holes. An attempt has been made to observe this radiation, which should bathe the Universe from primordial mini-holes, but no such radiation has yet been detected.

Inclusion of Matter distributed with uncertain wave nature does not seem to help in resolving the paradox of disappearing Matter at the centre of a black hole. What is worse, however, is that it brings along a new paradox of its own. For Einstein's General Theory of Relativity now becomes

Curvature of Geometry = Energy of Uncertain Matter Distribution

On the right-hand side is an uncertain quantity which has a probability of taking any particular value, such value varying according to the uncertainty of the Matter distribution. On the left is the certainty of geometry. As Einstein's original theory can be regarded as equating marble (from the elegance of the geometry) to wood (from the arbitrary, possibly gnarled, shapes of the Matter distribution), the new version above is equivalent to equating marble to sea spray. The spray floats all over the place, one moment there and the next not. But then how can marble ever have such a variable appearance? It is only possible if the marble itself became spray-like, by the gravity being uncertain, so that the marble-like curvature of gravity and the wood-like energy of the Matter would be equatable. The new version of Einstein's model therefore becomes

Curvature of Uncertain Geometry = Energy of Uncertain Matter

Both sides of this equation have the same quantum mechanical form. This revitalized version of Einstein's 1915 model is now ready for a test run.

In order to have a test run of any scientific model one has to work out predictions for the model from particular experiments. In the case of gravity this might be, for example, the motion of a planet round the sun, or the gravitational scattering of two 'elementary' particles, such as neutrons, off each other. If such predictions are attempted it is found, as when first tried in the late 1920s and early 1930s, that no sensible results can be obtained. All numerical predictions become infinite. So did those for processes that only involved Matter. The non-gravitational case was made finite by absorbing the irksome infinities in the predictions through a change of the masses and charges of the interacting particles. This so-called 'renormalization' gave excellent agreement with experiment. In the case of radioactivity, it was necessary to use the gauge principle (mentioned in Chapter 5). This paved the way to the unification of electromagnetism and radioactivity described briefly in that chapter. An analogous approach to the nuclear force led to chromodynamics, also now well tested. The renormalization programme of the 1950s and 1960s paved the way for the modern success in high-energy physics. It is even possible to express the equations of a theory which can be renormalized in such a way that no troublesome infinities need ever be seen.

This highly successful programme for particle physics did not work out for uncertain (or quantum) gravity. The latter is a theory which cannot be renormalized in the same fashion. There are an infinite number of different sorts of infinity which arise in trying to make experimental predictions, and there are only a finite number of particle masses and charges. Yet there seems to be no way to avoid facing up to the problems of constructing a sensible quantum gravity. How to proceed?

One approach is to bury one's head in the sand, like an ostrich, and hope the problem will go away. Some scientists have done that. A more constructive way forward is to claim that the specific techniques used to calculate experimental predictions in which gravity is involved are too weak and that these exasperating infinities would disappear when more powerful techniques (yet to be discovered) are used. Since no one knows what these latter approaches are, this avenue has not proved very useful. The third way has been to look for ways of ameliorating the infinities in the particle theories and then ascertaining if that helps the gravitationally extended theories. It is that route which has finally led to a finite theory of quantum gravity by way of what are called 'superstrings'. These involve the phenomenon called supersymmetry. Here we will take a breath and move on to a new section.

All things super

THE HUNT FOR finite interacting particle theories began in the early 1980s, after the introduction of a beautiful symmetry between all of the particles of nature. These latter come in two forms which can be divided into Matter proper (protons, neutrons, electrons, etc.) and radiation (photons, W and Z-mesons, gluons). Supersymmetry is a symmetry which transforms these two classes into each other:

Matter Proper ⟷ Radiation

The symmetry is therefore a very big one and indeed it can almost be regarded as the biggest one in particle physics; any other symmetry would have to combine Matter particles (as in the eightfold way) separately with each other and radiation particles separately with each other, or involve supersymmetry so as to cross the Matter–radiation gap.

A theory of particles will possess supersymmetry if it gives the same predictions independently of the specific choice of particles and radiation which can be combined together under supersymmetry. These combinations (like the eights and tens of the eightfold way) will be of finite size (and these so-called supermultiplets have been completely classified). It was soon realized that non-gravitational supersymmetric theories existed which were completely finite, although none of them appears to be realized in nature. In order to construct a gravitational theory which was supersymmetric it was necessary to construct the super partners of gravity itself. This was done and an extension of geometry to supergeometry was also achieved. However the resulting supergravity theory, although having a reduction of the nasty infinities, still had an infinite number of them which could not be renormalized away. This was the theory which Stephen Hawking had hailed several years before as 'the TOE'.

At this point the story changes, as does the value of the dimension of the Universe. A problem which had attracted attention in the early 1920s was how to unify gravity and the other forces of nature without worrying about uncertainty and quantum mechanics. If we accept that gravity is the curvature of geometry in the four dimensions of space–time, then in what dimensions are the other forces curvature? Curvature in an additional single spatial dimension was found to be equatable with electromagnetism. We know we do not live in four–space and one–time dimensions at our usual energies, but we cannot be sure that the extra space dimension beyond our usual three might not become revealed to us if we had enough energy. In the very early stages of the Big Bang, say, this further space dimension (to bring in electrodynamics as curvature along it) may have been as important as the others. At present energies in the Universe it might, for example, be curled up into a tight little circle, only observable later at the Big Crunch (if that comes). Presently the only relic of the extra fifth dimension would be the force of electrodynamics itself. This approach to unification by adding curled-up higher dimensions can also be made supersymmetric and for a while an eleven-dimensional version of supergravity became very popular (it being the largest dimension in which such a supergravity theory could be constructed). But it still did not allow sensible predictions to be made when a quantum version was constructed. By the early 1980s the way ahead seemed blocked.

Then, instead of considering point particles, it was proposed to analyse strings or loops moving in space–time. String theory had been around since the 1960s but had made relatively little progress. Now it was discovered how to include supersymmetry all around each loop, thereby constructing the superstring. Moreover it was found possible to remove all of the infinities for a very restricted solution, which could only exist consistently in ten dimensions (the non-super version required 26!). There were therefore six extra dimensions for curvature to give rise to the other forces of nature. The particles of nature arise as the transverse vibrations of the string – much like a drum's notes when hit. Gravity and electromagnetism can be shown to arise from the lowest modes of vibration.

This discovery sparked off an enormous explosion of interest in the subject, which has now abated as a result of two critical features. Firstly, it has proved extremely difficult to obtain predictions of experimental phenomena which allow the other forces of nature to be included correctly. The problem is more difficult than that for the nuclear force, which itself has so far resisted all attempts at solution. Secondly, there is as yet no elegant formulation of the (now many) candidate superstring theories along the lines of Einstein's beautiful geometric approach to gravity. The natural framework would be that of the geometry of loops in superspace, but that has not yet been possible to construct. A very special approach (called the light-cone gauge) has had to be developed to verify that the predictions are sensible (finite) and that has only been possible for one special ten-dimensional superstring called the 'heterotic' (a mixture of two sorts of strings). Other superstring versions have still to be analysed and undoubtedly some very hard mathematical problems lie ahead. These difficulties have caused a drift away from the superstring itself in order to look at related topics.

My conclusion is that at least one superstring has been proved to give sensible physical results, but that a lot more work must be done to show that this, or any related superstring theory, fits the facts of nature. It is clear, from the existence of this single sensible quantum theory of gravity, albeit in ten dimensions, that the only way forward into uncertain gravity is through loop spaces, very likely in higher dimensions. Arguments based on applying the standard methods of quantization directly to Einstein's original theory cannot be trusted for any sensible understanding of the Universe. In particular the transverse vibrations of the superstring, whose lowest modes are expected to be related to the quarks etc., which we described in the previous chapter, will become ever more important at the higher energies towards the beginning of the Big Bang or as Matter disappears inside a black hole. Thus although we do not yet have a solution to the above paradoxes of the appearance or disappearance of Matter *ex* or *in nihilo* we do expect the framework of superstrings to allow us to discover if the use of uncertain gravity allows a resolution of the paradoxes.

Finally, where is the TOE so gleefully heralded by the scientific optimists? First it was eleven-dimensional supergravity. When that was seen to fail to give sensible answers, the heterotic superstring was installed as the most likely candidate. Enormous claims have been made about the superstring as the TOE. In my opinion, it will be at least several decades before we know if superstrings can include even the known forces of nature, let alone give an explanation of cosmological features. It is the problems raised by the latter which we consider in the next chapter.

Chapter Seven

THE BEGINNING OF THE UNIVERSE

The hot Big Bang

FOR MILLENNIA the starry firmament has been the subject of poetical description and philosophical debate. It is only relatively recently that scientific analysis has been brought to bear on the nature of that glittering array. Bigger and better telescopes have clarified the structure of our own Milky Way and its place among the myriad of similar spiral or elliptically shaped galaxies and allowed increasingly sophisticated theories of the structure of the Universe to be tested and refined. In the last decade there has even been a joining together of our knowledge of the sub-atomic world and the cosmos. Particle physics has turned out to have crucial implications for certain aspects of the nature of the Universe and this has been an even greater spur in the attempt to unify all of the forces of nature.

Modern cosmology is based on observation, as is all science. Exploration of the solar system and the Milky Way was first initiated by Galileo in 1609. His discovery of the four satellites of Jupiter lessened the likelihood that the earth was the centre of the Universe and validated the idea of Copernicus that the earth moved round the sun (and for which 'heresy' Giordano Bruno was burnt alive by the Catholic Church). That there were other 'universes' (which we now realize are other galaxies) outside our own Milky Way was not proven until large telescopes were built which showed details of the beautiful internal structure of the cloud-like objects, or nebulae, first seen in the latter part of the eighteenth century. It was not until the early part of this century that the controversy over the distance from our own solar system to these nebulae was resolved.

It was the American astronomer Edwin Hubble, mentioned briefly in Chapter 1, who clarified this point. There are variable stars (whose brightness changes periodically) that can be separately seen in some galaxies relatively close to our own. Knowing that these variable stars have a certain brightness, the intensity with which they appear to us allows us to calculate how far away they actually are. In this way, Hubble confirmed convincingly as an example that the galaxies unimaginatively named M31, M33 and NGC6822 were definitely island universes far outside our own galaxy.

He continued this work for other galaxies, at the same time trying to measure the speed with which these galaxies were moving with respect to our own. To do this he used the Doppler effect, in which light emitted by a galaxy moving away from us looks more red, or towards us more blue, than if the galaxy were at rest (as occurs in a car siren or train whistle dropping its frequency as it passes us). Hubble found that the galaxies were all streaming away from us: the Universe was exploding. This has been followed up by numerous further observations, all of which agree with the basic conclusion that the Universe seems to have exploded from a highly dense primordial blob about fifteen billion years ago. This was the so-called 'Big-Bang' (although it could not have been heard because there was no 'outside' to it). Some of the most distant galaxies are now moving away from us at a speed close to that of light with respect to us; when they reach the speed of light we will no longer see them and they will have passed beyond our aptly named 'event horizon'. If the expansion of the Universe, discovered by Hubble, continues indefinitely, all but the group of galaxies in our locally bound cluster will end up speeding away from us so fast that we will not be able to see them. More careful observation may, on the other hand, show us that if the headlong rush of far-off galaxies away from us is slowing down it might ultimately reverse to end in the 'Big Crunch', with all the galaxies collapsing back on themselves.

The ultimate fate of the Universe depends, then, on further evidence. If continued expansion occurs, space travel in the future will be limited in terms of the number of places to visit, because there will be fewer galaxies inside our event horizon – and we cannot ever travel beyond that. On the other hand, the nature of the Big Crunch brings us face to face with problems as difficult as those we have to solve regarding the expanding universe, or of Matter disappearing at the centre of a black hole: how can it be done without destroying the laws of physics in which we believe, and in what way must scientists modify these laws to obtain a sensible result? It seems that the ultimate fate of the Universe could lead us into the same minefield of difficulties as did the beginning of the Universe and the centre of a black hole.

The Big Bang requires that at some time in the past the Universe was suddenly created. After that it expanded in a manner that can be explored in detail. But before that time there was no Universe – and no time! One way to avoid the enormous difficulties presented by the Big Bang is to deny it happened. This was the approach taken in the early 1950s by a group of Cambridge theoretical astronomers in their theory of 'continuous creation'. It was suggested that Matter was being continuously formed to fill up gaps left by the expansion. In this way, the Universe could have existed for ever (allowing the name 'steady state' for the new theory) and the problem of the unique creation of the Universe was neatly sidestepped. However in the 1960s a new phenomenon, 'the microwave background radiation', was observed which indicated that the Universe started off at high temperature. If this had been so, as it cooled there should now be a Universal relic radiation of the hot early period, which would have a temperature corresponding to our present cool epoch. This radiation (with a temperature of about three degrees above absolute zero), was discovered in the mid-1960s at Bell Laboratories in the United States.

Besides the discovery of the microwave background radiation, further strong support for the hot Big Bang has come from the relative proportions of elements – hydrogen, helium, lithium – observed overall in the galaxies (and measured by the characteristic radiation they emit). These various sorts of Matter could have been produced in the very early hot stages of the expanding Universe by nuclear reactions. Detailed calculation of these processes were first carried out in the 1960s, and more refined ones quite recently, and led to excellent agreement with the observed amounts (helium being about a quarter as abundant as hydrogen, for example).

These observations have all been incorporated into what has been called the 'standard model' of the Universe. To the observed isotropy of the distribution of Matter (observationally and in terms of the microwave radiation) has been added the homogeneity of the distribution of the galaxies. There are local clumps and voids, but there is no special place that can be singled out and where it can be claimed 'Here is the centre – this is where the expansion began.' The standard model uses the theory of gravity – general relativity – as proposed by Einstein in 1915 and discussed in the previous chapter. The Matter is taken simply as a uniform distribution of some sort of dust. The equations of the theory can be written in a simple form when homogeneity and isotropy are assumed. They involve crucially an overall scale factor, indicating the mean distance between the dust particles. It is this scale factor which determines the scale of the Universe, since if it is doubled then the size of the Universe would also double. The scale factor is zero at the Big Bang itself, since the Universe had no size then, and increases in a way depending explicitly on whether or not the Universe is infinite in extent (like a flat plane going on for ever) or is finite (like the surface of a sphere – which is bounded in area with no tangible boundary). The scale factor also allows us to calculate how the Universe cools as it expands.

The resulting scenario can be explored theoretically back to the unimaginably short time of about a billion billion billion billionth of a second after the Big Bang. This can be achieved by including in the description of Matter various of the particle physics versions described in Chapter 5. One such is the standard model of electro-weak and chromodynamic forces. Matter would be composed of the three families of particles described in Chapter 5. At the hottest early stage after the Big Bang, the forces of nature are expected, on theoretical grounds, to be more unified and symmetrical than when later cooling occurred. Thus, after a ten thousand billionth of a second, the electro-weak unification would have ceased to be evident and electromagnetism and radioactivity would have been experienced as separate forces. The temperature would have then dropped to about a million billion degrees Centigrade. At about a millionth of a second the quarks would have condensed into protons and neutrons; attempts to glimpse these quarks in very hot matter are currently being made when intense high-energy beams of heavy nuclei are focused on targets here on earth. If they are found, then there is increasing scientific expectation that the hotter the Universe became the more symmetrical and unified would be the forces governing it. At about a hundredth of a second after the start of the Universe, particles such as protons would have combined with their so-called anti-particle companions (which have opposite electric charge and other features, but identical mass). Just about three or four minutes after the beginning (at about a billion degrees Centigrade) various isotopes of helium nuclei would have formed by nucleosynthesis, as mentioned earlier. Moreover no more than three families are allowed in this scenario without changing the ratios of the elements formed beyond that observed. At about a million years, and about three thousand degrees Centigrade, atoms would finally have formed, the electrons being sufficiently cool to remain close to the nuclei attracting them. The Universe would now become transparent to light; before this, all the electrons were free to scatter the light, causing the Universe to be opaque. The microwave background radiation would have de-coupled from Matter and cooled down ultimately to the 2.7 degrees we observe today. Finally, at about a billion years after the Big Bang, galaxies would have formed.

The standard Big Bang model is remarkably successful, from which we conclude that the Big Bang has strong support. Its occurrence explains a range of observed phenomena in terms of events which took place over an enormous range of times (a hundredth of a second up to fifteen billion years, as I described above). The model dovetails nicely with the enormously successful particle physics developments, especially the agreement on only having three families of truly elementary particles. What more needs to be said?

Inflation

As we concluded in the previous section, hot Big Bang cosmology is in good shape. There appears to be no viable alternative. There are still some aspects, however, to be filled out. The most important of these is that of structure in the Universe. There are many details of this structure that have been seen when the distribution of galaxies is considered: clusters, superclusters, voids, walls, as well as smaller-scale features such as the detailed shapes of galaxies themselves, of stars and of planets. In this arena there are heated controversies raging as to how such structures arose. It is not the underlying framework of hot Big Bang cosmology that is at stake, but the detailed solution of the structure problem itself. Since this can have relevance to the final fate of the Universe – Big Crunch or ultimate expansion – I feel it is important enough to consider here even though the answers are very unclear.

The standard cosmological model has, in fact, no hint of producing any structure in the Universe. If we go back to that early time when Matter and radiation in the Universe did not affect each other, the radiation itself could have been released from places in space which at that time were outside the range of influence of each other. It is clearly a difficult problem to explain how any two such places, so effectively causally unconnected, could be at almost exactly the same temperature at the same time and so lead to the observed common background radiation and distribution of Matter.

About a decade ago, in order to obtain a resolution of these and related difficulties, it was suggested that at a very early period the Universe suddenly expanded at an ever-increasing rate, depending on time in a so-called exponential manner. This growth is similar to that which occurs in inflation, where 100 per cent corresponds to doubling of prices every year.

This 'inflationary' expansion easily solves the problem we mentioned above, since it enlarges the regions of causal influence in different parts of the Universe by a very big factor very rapidly (depending, of course, on the amount of inflation). Thus one region which started out much smaller than a proton ended up an instant later the size of a grapefruit and later became our own Universe. It also explains how the Universe could appear 'flat' – with parallel lines never meeting – by having just the right amount of Matter (the so-called 'critical density') to balance between the Big Crunch and exponential expansion. This looks a rather unlikely scenario without inflation, but is easy to understand with it, just as blowing a beach ball to one thousand times its normal size would make its surface appear much flatter. It is important to note that such a type of expansion (as if gravity became highly repulsive) can be shown to have its natural origin in the sort of particle theories presently in use in the electro-weak theory described in Chapter 5. Such expansion arises as an offshoot of a special mechanism for making masses of all the particles in the electro-weak theory. This uses the energy of a single further particle which is avidly being sought for in the biggest particle accelerators; its discovery would be a very strong justification of the inflationary scenario in hot Big Bang cosmology.

Let us return to the problem which I raised at the beginning of this section: the origin of structure in the Universe. As noted in Chapter 5, a distribution of energy will have constant fluctuations in its intensity at the subatomic level due to quantum effects, just as waves dance up and down on the surface of a lake. The peaks created by these fluctuations would, after inflating, become large enough to serve as seeds of stars and galaxies. Further clumping would occur naturally under the action of the gravitational attraction between the initial clumps.

All this looks very successful indeed, except for one big question. The inflationary theory needs enough Matter to reach the critical level to have a'flat' Universe. Yet only about one per cent of that required has been seen by visible light. Where has the missing Matter gone? Studies of the motions of stars with galaxies have shown that these galaxies may contain ten times more invisible than visible matter (say as black dwarfs – dead stars – neutron stars or black holes). A similar result is arrived at by observing how individual galaxies are moving inside a cluster of their fellows. But that still leaves 90 per cent of the Matter to be found for inflation to work properly. However, the calculations about the synthesis of helium and heavier nuclei in the early stages of the Big Bang talked about earlier also would not work if there were more protons and neutrons than observed. So even if there existed extra material above and beyond the ordinary dark Matter in galaxies, it cannot be in the form we presently know of. There has to be some new form of Matter, making up about 90 per cent of the total material of the Universe. It is needed in order to let inflation lead to the properties of the observed Universe. Yet this new form of Matter can only interact very weakly with ordinary material, since otherwise it would have spoilt the early stage of nucleosynthesis. Some scientists, tongue-in-cheek, have coined the acronym 'WIMPs' for this putative new form of Matter, meaning Weakly Interacting Massive Particles. They have not yet been observed.

There are two distinct suggestions for the form of the invisible Matter – either 'hot' or 'cold'. The hot version, involving fast-moving particles identified as neutrinos, have two problems. Firstly, neutrinos have little or no mass and certainly not enough to brake the expansion of the Universe. Secondly, computer simulations show that fast-moving neutrinos would have taken too long to settle into galaxies. The alternate cold, dark Matter theory (involving slow-moving particles) is now the favourite, although this also has difficulties. One is that nobody has observed any particles as candidates to constitute the necessary extra material. However, that is not insuperable; in the unified theories there are various particles with just the right properties to fit the bill (light super companions or extra particles of grand unified theories). The search for such particles here on earth, and the excitement engendered by such a search, is an indication of the amazing progress made in relating the science of the very small (particles) to the very large (astronomy). That is, of course, completely to be expected when the very early Universe is being analysed, since the crucial distances involved are all down to particle physics sizes.

A second potential difficulty with the cold invisible Matter plus inflation idea is that it is expected to lead to the presence of small but detectable inhomogeneities in the microwave background radiation. However, these have very recently been observed at exactly the level expected by the Cosmic Background Explorer satellite experiment. This has been hailed as an important experimental support for the theory. These effects are thought to have been generated by quantum fluctuations at a minuscule time after the Big Bang and, due to inflation, produced disturbances in the background radiation which are now thousands of millions of light years across.

The third (and what seems most serious) difficulty is concerned with intermediate-scale structure, associated with clusters of galaxies, and the presence of voids and walls in the distribution of galaxies. This problem was particularly brought to the fore by results from the Infrared Astronomical Satellite (IRAS) in January 1991, which revealed galaxy clumping on a scale too large to be accounted for by the usual version of the cold invisible Matter model. However, modified versions of the theory, it is claimed, can explain the observations.

There are still further puzzles about cosmology to be resolved. I turn to them next.

In the beginning

IN THE PREVIOUS two sections I described the slow but sure development of a cosmological theory probing back ever further towards the beginning of the hot Big Bang. More and more sophisticated experiments are needed, using both satellites and bigger telescopes, to look at the firmament, and ever larger particle accelerators observing deeper into Matter here on earth. This will allow decisions to be made between various opposing theories. Such progress is expected to continue indefinitely, or at least until the scientific community fails to rustle up any more financial support to build the next generation of equipment. However, the models being refined always possess the same general form. They allow us to predict what the Universe will be like at a later stage if it is assumed to have been in a particular state at an earlier one. As the earlier time is pushed back further and further it appears possible to conceive of arriving at a need to define the initial conditions of the Universe at the moment when it was first created. At no time would we be able to justify these initial conditions, except in terms of their success in predicting the observable details of the ensuing evolution of the Universe.

The picture of the Universe being created suddenly in an already existent space–time framework is fatally flawed. For there is no way for any theory to select a particular time at which the act of creation could occur. Nor is a particular place of creation of the Universe able to be specified. Only the probability of creation in any given time period or spatial volume is able to be predicted. But then multiple creations would occur, with ensuing collisions. None of these have been observed; the model of a fixed background space–time must be abandoned. One can try to inflate the space–time of these separate Universes so they do not collide, but this seems to be very *ad hoc* and has not received much support. Space and time are also necessary elements to be created at the same time as the Matter, and not to have been assumed *ab initio*.

In any case, part of the framework of general relativity is that Matter and space–time are inextricably interlinked and we would not like to bring about a divorce at this stage of the proceedings.

The problem we are faced with, then, is to decide on a choice of the initial conditions at the beginning of the Universe for the gravitational field and Matter distribution, so that the resulting evolution produces the observable plethora of planets, stars and galaxies and their clusters. This has been attempted by numerous scientists but it is now accepted that it is impossible to choose suitable conditions. One evident reason for this is that in the beginning the gravitational field strength was infinite; that is the very nature of the Big Bang – which started with the Universe having no size at all. It would be necessary to describe in specific detail how the gravitational field of the Universe increased as one turned the clock back and got closer and closer to zero. This is a very difficult numerical problem. The second source of difficulty is the evident problem of trying to deduce the enormous complexity of the evolving universe from the delineation of its very initial state. That also is thought to be impossible to achieve. A third and crucial reason is that in any case the simple classical approach to cosmology is expected to break down at very high temperatures, especially those that would occur at the earliest stages of the hot Big Bang. This is because quantum effects become important. We already relied on quantum mechanics in the calculations that were used for nucleo-synthesis at quite an early stage, but the gravitational field was not itself quantized. However at an early enough time the quantum nature of gravity will become evident, since the gravitational field will itself have become as important as the other forces, whose quantum nature is unassailable from an experimental viewpoint. In that situation there can be expected to be a whole set of possible gravitational and other fields, each with a certain probability.

The initial conditions are now changed from criteria selecting a particular initial gravitational field to a method of choosing a probability for choices of possible 'initial' gravitational fields. That does not seem all that much of an improvement. However, because it is a more complex problem there is a better chance that one may explain the complexity of the observed Universe by means of it. Thus the situation may have improved.

Quantum creation of the universe

THE BASIC PROBLEM being discussed here requires the use of a special concept. The choice of a probability for each possible position of an electron corresponds to defining what is usually called the 'wave function' for that electron. We can liken a wave function to a landscape spread over the values of the electron's position. The landscape has hills and valleys whose height or depth determine the probability of the electron being at a particular place. Thus a very high peak indicates a very strong likelihood that the electron is there. We can similarly introduce a wave function depending not on an electron's position but on the initial shape (or geometry) of the Universe. This will tell us how probable it was that the Universe had a particular shape. The question at issue, then, is to try to determine the wave function for these shapes, or what may be termed in shorthand the 'wave function of the Universe'. We have, moreover, to clarify the notion of time itself in this context. As I have noted earlier, both time and space are regarded as being created in the same event as the Matter in the Universe. We cannot therefore talk about conditions selecting the 'initial' wave function for the Universe. What must be attempted is to discover a condition which uniquely selects a wave function for the Universe and then try to interpret one part of the shape on which this wave function depends as a time, with some suitable properties. One such might be, as we experience in our own time, that only positive energies are present for particles moving into the 'future'.

There have been numerous attempts at finding a suitable criterion along the above lines for the wave function of the Universe. I have already mentioned one by Stephen Hawking and his American colleague Jim Hartle. It involves defining such a wave function in terms of the ways in which the three-dimensional shapes on which it depends can be joined smoothly on to four-dimensional ones. This can be done in a mathematically precise manner provided that the fourth dimension involved is purely imaginary. This makes the space and time variables precisely equivalent. The condition can thereby be interpreted as requiring that there can be no initial point in time, in the same manner that there is no initial point in space. The results of the criterion cannot be evaluated for any realistic gravitational theory, such as Einstein's, because of the inherent difficulties in a quantum version of Einstein's theory (as mentioned in Chapter 6). However, it seems to give valuable insights in 'toy' models with enormous reduction in the number of quantities important in describing the gravitational field.

An alternative criterion, also described briefly earlier and without the use of imaginary time, required that the wave function of the Universe should vanish (in a suitable way) on shapes (gravitational fields) which are infinite. This is evidently a way to avoid the gravitational singularities, such as arise at the centre of a black hole, which were so hard to answer from a full quantum gravity. Yet again, this definition does not make sense with respect to Einstein's theory of gravity. It does seem to make some sense for highly simplified models of gravity and Matter – 'toy models' – as in the Hawking–Hartle prescription. Yet they are both infinitely remote from any real test. Any claim to have explained the beginning of the Universe thereby is unacceptable. Moreover, they are prone to more technical difficulties associated with the nature of time.

One might try to extend the discussion to the more hopeful superstring theory mentioned earlier. This is expected to have a very different structure from that of Einstein's gravity since it involves an infinite tower of increasingly massive particles (the transverse string vibrations) corresponding to companions of the gravitational field. Increasing numbers of particles in this tower are expected to be noticeable as the temperature increases in going back more and more closely to the Big Bang itself. These should change the results considerably. Whatever final structure ultimately ensues, it is expected to be very different from that for Einstein's gravity. The wave function of the Universe would then have to be selected as a choice of probability for the set of string fields. This has yet to be done, and looks a formidable task to analyse. It may become simpler if a more elegant formulation of superstrings is discovered. Until then the idea of quantum creation of the Universe is an interesting possibility but must be admitted to be completely unproven.

No beginning?

THE PROBLEMS FACING the quantum approach to creation of the Universe which were raised in the previous section are highly important. Moreover there is always the question as to how a particular theory (such as quantum mechanics, or superstring theory) came to be arrived at in the first place, even if a criterion for the correct wave function of the Universe can be discovered in terms of this theory. It would also be natural, in the face of the difficulty of 'why that criterion on the wave function of the Universe?' as well as 'why that theory?' seriously to consider the case of an infinite regress of theories and wave-function criteria. Indeed the models of the previous section might be regarded as attempts to discover preliminary wave functions produced on the basis of a 'halfway house' theory, one not being full quantum gravity but possessing some of its characteristics. The next steps would be to take the full quantum gravity, with its criteria for a wave function for the Universe. But why stop there? The whole search for initial conditions then takes on a completely different complexion.

Let us admit that there is an infinite sequence of theories, each one being reasonably precise over a range of energies, but breaking down above a certain energy. This happens for classical physics in that quantum physics takes over when atoms become broken up and atomic structure must be used in describing properties of the systems. Let us apply this to the case of the expansion of the Universe in the hot Big Bang. As cooling occurs, one theory needed to describe the Universe can be replaced by another which is more appropriate to the lower energy range. The initial conditions for this lower energy theory would be expected to be determined by the final state of the higher energy theory. That final state would itself be determined by an initial state of that theory, evaluated at the energy at which this theory just became appropriate. That initial state would then be determined by the final state of the next higher level theory, itself valid at the next higher range of energies. And so it goes, each theory determining both the structure of the next one and also its initial state.

We had earlier proposed this infinite regress of theories as a way to resolve the impossible problem as to 'why that TOE?' but still preserve causality. Here we propose an extension of this infinite regress as a way of avoiding the similar question 'why that criterion for the wave function of the Universe?' We do not wish to jettison any of the ideas of elegance to be used in our search for suitable wave functions. However, the new framework of the infinite regress takes the pressure off the scientist from trying to play God. He is now, in his search for better theories and their wave functions to describe the Universe, only performing his natural function of searching, conjecturing, testing and searching further, etc. However, he should never regard his discoveries as more than steps along the infinitely long road to perfect knowledge. There is no ultimate certainty at any stage.

For an infinite regress of this sort to exist there must be an indefinite number of critical experiments ahead. Each one will destroy the theory valid up to that level and require it to be replaced by a more effective and more general one. It would seem to be impossible to describe completely the creation of a Universe, governed by an infinite regress of scientific laws, by singling out the scientific theory at a particular level. For in so doing the Universe will be lacking the right properties of the infinite set of higher levels. Thus although the idea of quantum creation of the Universe might help us to make further progress in understanding the various levels, it will never explain the existence of the

Universe. Evolution of the Universe is occurring successively at ever lower levels as the Universe expands, according to the infinite regress approach. But since there can never be a complete description there still can be no quantum creation of the whole Universe. It can be regarded as uncreatable, at least in our understanding, since it very likely has infinite complexity. But that infinite complexity is explicable only with respect to a precise infinite regress of scientific theories, each having its own individual structure but being more complex than the one at the next lower level.

The infinitely regressing Universe can therefore be looked at in one of two equivalent ways. In one scenario it was constantly being created as the Universe expanded, as one level of theory and explanation descended to the next. In the other the Universe was always here. It had no beginning. Nor can it have an end, since in the Big Crunch (if it occurs) we would expect there to be a reversal of the history of its earlier expansion phase. Thus the Big Crunch would never be reached, although humanity, the solar system, all galaxies and all structure would successively disappear as the temperature rose. Increasing symmetry would reappear and more and more complex mathematical structures would be necessary to describe the increasingly complex nature of reality as ever deeper levels became apparent.

PART THREE

THE MENTAL UNIVERSE

Chapter Eight

THE MECHANICS OF THE MIND

The evolution of life

IN ORDER TO preserve the possibility of continuity between differing levels in the Universe – atoms and molecules, nuclei, electrons and atoms, etc. – that has been successful over the whole range of the material world, we must also show continuity between non-living and living things on earth. That life arose from non-living material was more or less accepted some centuries ago by the scientific community, but the manner in which that occurred has been far more controversial.

The two extreme positions taken were that it occurred either by spontaneous creation or by complex evolution. In the former, living forms such as intestinal worms or microbes were supposed to arise by chance, the more appropriate the external conditions the greater the probability of life being thereby engendered. The opposing complex evolutionary case was succinctly posed in a 1912 address to the British Association for the Advancement of Science:

> So far from expecting a sudden leap from an inorganic, or at least unorganized, into an organic and organized condition, from an entirely inanimate substance to a completely animate

> state of being, should we not rather expect a gradual procession of changes from inorganic to organic matter, through stages of gradually increasing complexity until material which can be termed as living is attained.

The final scientific abandonment of the spontaneous creation approach can roughly be dated from the appearance in 1936 of the monograph *The Origin of Life* by a Russian biochemist. In this book the innate complexity of living things was emphasized. It was inconceivable that they 'could appear in a very short time, before our eyes, so to speak, from unorganized solutions of organic substances'.

The early conditions on the surface of the earth in which primitive life first evolved, in terms of both the material available and the energy coming from the sun, are reasonably well understood. The atmosphere would initially have consisted mainly of hydrogen, with possibly some methane and ammonia. Carbon dioxide from volcanic gases then replaced the methane. A third stage arose as oxygen began to be produced by photosynthesis from early living plant cells. There is good evidence that until these cells had achieved a highly organized and protected state, free atmospheric oxygen would have been dangerous since its presence would have rapidly oxidized any precursors of life and prevented their existence and subsequent evolution. Oxygen has a deleterious effect on contemporary cell nuclei, which indicates that anoxygenic conditions may have prevailed during the early evolution of the cell. It was not until about two billion years ago that the rate of oxygen production by photosynthesis exceeded the rate needed to oxidize volcanic gases, allowing oxygen build-up to occur over the earth's surface. It is thought that the sun's ultraviolet light, able to penetrate the earth's atmosphere before the formation of the ozone layer, was the most important primitive energy source acting on the 'warm little pond' of ammonia and phosphoric salts, in the words of Charles Darwin. Other sources of energy were derived from electrical discharges and possibly from decay of radioactive material in the earth's crust.

One of the first scientific experiments carried out showed how amino acids, the building blocks of proteins, could be synthesized from mixtures of methane, ammonia, water and hydrogen. An electric discharge acting on such a mixture for about a week produced various amino acids; up to fourteen of the twenty common amino acids found in proteins are claimed to have been synthesized in this manner. The crucial base pairs in DNA were later synthesized by various groups. Hydrocarbons and fatty acids were also formed by electric discharges in methane, and large organic molecules could have been built up by dehydration of mixtures of amino acids; this process could have occurred on clays near the shoreline under the heat of the sun.

The final, and possibly most important, step is to discover how nucleic acids were formed so that they were then able to control protein synthesis in the highly specific manner observed in all living systems. This is presently a very active field, but one in which the final answers have still to be elucidated. As a researcher in the field wrote recently,

> We are optimistic that the path of chemical evolution will be outlined in the laboratory. The biochemical knowledge which has been amassed within a few years has given us a deep insight into some of nature's most secret processes. With this understanding to help us, the time needed to solve our problem may not be long.

Traces of life have been found in rocks three and a half billion years old. For example, it has been reported that stromatolites (organo-sedimentary structures resulting from the growth and metabolic activity of micro-organisms) have been found in rocks in Western Australia and Southern Africa which could be as much as 3400–3500 million years old. Organically walled microfossils showing stages of cell division have been reported in a stratigraphic unit (a layer of rock laid down at the same time) over 3400 million years ago in Swaziland. There has been controversy over the presence of biogenic material in the oldest known terrestrial rocks, dating from about 3800 million years ago, only some 800 million years after the generally accepted age of the

Earth. Some opinion holds that life could even have started 4000 million years ago.

From that early date life has evolved into a wide variety of forms here on earth, in a process first appreciated by Charles Darwin and Alfred Wallace in 1858, on the basis of species variations. Evidence for evolution has now come from many sources, the main ones being palaeontology (the study of fossils), geographical distribution of species, classification, plant and animal breeding, and comparative anatomy, embryology and biochemistry. Neo-Darwinian evolutionary theory, as it is now called, is based on evidence from a broad range of these sources and is supported by a mass of otherwise unrelated observations. Evolution is widely accepted among scientists, but there is still much to be done in refining the theory and its application to all observed situations. Darwin's final paragraph in his *magnum opus*, *On the Origin of Species*, aptly describes the perspective that the evolutionary theory grants on the world:

> There is a grandeur in this view of life, with its several powers, having been originally breathed by the Creator into a few forms or into one; and that, whilst this planet has gone cycling on according to the fixed law of gravity, from so simple a beginning endless forms most beautiful and wonderful have been and are evolving.

We can now begin to appreciate the power of the laws of nature in leading from the Big Bang to the wondrous varieties of life here on earth. Who knows what other varieties of life might exist on other planets around other stars, in our own or other galaxies?

Now we must turn in detail to the second theme of this book – the mental world. When did 'Mind' appear on earth? From the strict materialist view it might have been expected to have evolved as part of the complexity of living systems increased. It is important to try to detect at what stage Mind can be recognized. I turn to this in the next section.

The evolving Mind

IT HAS BEEN said that there are over a million living species here on earth. In spite of the enormously disparate range of shapes and sizes these come in – from the lowly amoeba to the elegant giraffe, from the blood-sucking leech to the soaring eagle – their variety has been explained by means of the evolutionary theory of Darwin and Wallace and by the more detailed DNA/RNA molecular basis of inheritance in terms of the genetic code, uncovered by Francis Crick and J. D. Watson in 1964. How genes control the development of bodily structure is now an established part of science and has led to the important and ambitious human genome project aimed at mapping out the complete set of genes in the human cell. This project is being carried out by a concerted group of scientists worldwide and is expected to take about a decade to complete. Genetic engineering, although somewhat controversial, has already produced new animal forms, as well as helping to cure some special forms of human disease.

Hand in hand with the variety of animal shapes goes the enormous range of behaviour patterns. Zoos hold a strong fascination for us in that we see some of these patterns in action. But what we do observe is only the tip of an enormous iceberg. We miss the amoeba in its search for chemical nutrients and the leech passing through its life cycle. The stickleback building a nest in the sand by sucking sand into its mouth and then spitting it out elsewhere is hidden from us. The ability of the chimpanzee to solve simple problems and possibly even to use sign language is also not something we normally observe at first hand.

The human-like level of the great apes was brought home to me most forcibly some years ago. I was involved in a science television programme for children in which a chimpanzee of about four years (over half grown) was appearing. We all (including the chimpanzee) took a break for lunch in the studio canteen. I sat opposite her and had a useful discussion with her owner as I ate my main course. I then turned to peel the orange on my dessert plate. She immediately held out the palm of her hand towards me and looked at me beseechingly. What could I do but share the orange, segment for segment, with her? She ate it very gratefully, each time putting out a hand for another segment. When I had finished she jumped contentedly into the lap of her owner and settled down for a well-earned rest! Of course, this is only an extension of the way monkeys in zoo cages react to visitors with titbits but this chimpanzee was human-like in her response.

A visit to the primate house in the zoo shows that chimpanzees are but the most intelligent end of a continuum – orang-utans, gorillas, monkeys and lemurs. Such gradual changes of intelligence clearly exist in all species as their evolution is traced. If we consider the evolutionary sequence

fishes → amphibians → reptiles → mammals

we see emphasized more clearly this change in powers of behavioural response to a changing environment, an important feature of intelligence. At the lowest end of this sequence, behaviour is stereotyped, although it can be complex. Thus the 'displacement activity' of the stickleback when it builds its nest arises as a bizarre response when the drives of aggression and fear are equally balanced. This is only one of many examples of the intricate sets of activities possible with a number of underlying fixed responses. As one climbs the sequence the behaviour patterns become richer until finally the human emerges as the ultimate triumph in complex actions.

The sources of this richness, occurring well before mankind on the evolutionary tree, can only be ascribed to the powers granted to their possessors by their brains. It is very likely that the chimpanzee I had lunch with had a mental representation of the significance of an orange somewhere in her brain. The visual stimulus of the orange must have elicited a past memory of the pleasure eating an orange gave her, which caused her to hold out her hand to repeat the pleasure. She was goal driven, the goal being to eat the orange, and the route by which she could achieve it was by looking at me appealingly and holding out her hand in a way I could not refuse. But such internal representations occur further down the animal tree. It has even been suggested that chimpanzees have a 'theory of Mind'. In the case of chimpanzees being experimented with for sign language it was found that they had the ability to predict the behaviour of their human caretakers according to the styles of behaviour of the latter. It was as if the animals in some sense had a theory about the minds of their caretakers. It has been observed that only chimpanzees and orang-utans can learn the significance of their own image as themselves when faced with a mirror. Gorillas have failed on this task, although there may be a version of the problem more suited to their own make-up and abilities in which they would succeed. Lower down the evolutionary tree this self-image seems entirely lacking as, for example, in dogs or cats. This does not mean that these animals cannot have mental representations of the outside world stored in their brains.

As I write this, I have been disturbed by the barking of my own dog. She was trying to get into an upstairs room in which she thought my wife (whom she follows around very closely) was sitting. At that point my wife walked up the stairs and called to her. The look on the dog's face, and her whole bodily response, was one of complete surprise and disbelief. It was as if she was saying 'But you are in that room!' She clearly had the mental representation that my wife was in there and that was why she was barking – to be let in.

One can go on further down the evolutionary tree. Rats, for example, are excellent laboratory animals since they can be trained in a matter of a few days to learn to negotiate a maze and to discover where there are retrievable rewards. They have been found to have mental representations of a maze which, if the cues in the environment external to the maze are rotated, rotate with the cues. It has even been found that there are single nerve cells – appropriately called 'place' cells – that respond only when the rat is in a certain place in a given maze. Still further down the evolutionary tree, insects learn to find food by cues. They even communicate to others of their kind where the food might be, as for example in the case of the bee. The lowly ant can learn the cues of food sources, and if those cues are moved the ant will be misguided unless other cues allow him to stay on the correct track. What these lowly animals can achieve with very small nervous systems, sometimes only composed of clumps or ganglia acting autonomously, is indeed remarkable.

All these behaviour patterns arise from processing by the nervous system at different levels. A headless chicken may be able to continue walking, under the influence of brain-stem reflex action, but will have no ability to guide itself. Brain-damaged babies grow up to be intellectually deficient and need special treatment to enable them to be independent in even simple tasks such as toilet training and feeding habits. Increasing numbers of cases of human brain damage have arisen from recent wars, from the rising number of car crashes and from disease, especially in older people. These, though sad to relate, have brought about enormous strides in understanding the subtle manner in which the brain seemingly controls all aspects of behaviour. This has been allied to a range of experiments on animals, such as the rats mentioned above, and also on cats and monkeys. But it is not only such advanced animals that can help us increase our comprehension of the physical control of the mental world, and so alleviate human suffering. The goldfish retina, for example, is apparently a close analogue to our own (the carp is better but more expensive to buy). A group of researchers at a hospital in Amsterdam are experimenting with goldfish to understand the nature of colour vision (which the goldfish possesses) and so help those humans with colour blindness. At the same time they are working with chimpanzees to understand better the way in which behavioural responses correlate with brain waves recorded from the surface of the scalp by electrodes (EEGs).

One way to study the evolution of Mind is the one we have just been discussing, that is, the evolution of brain. It assumes that mental events are determined preponderantly, if not completely, by brain events. As such, the evolution of Mind involves tracing the details of the way in which brains of animals have evolved as they ascended the evolutionary tree, and how the concomitant increased range of behaviour patterns can be explained. In the process it is possible to understand how a particular response or behaviour pattern came to be modified by the development of a particular brain region. For example, even the way that the body is represented in the sensory part of the brain varies with species. The pig has a very large sensory brain region devoted to its snout, while the human has an exaggerated consignment of brain tissue for the face and hands. There are many patterns of behaviour of animals which can be seen to be correlated, in a somewhat general fashion, with the brain they possess. By understanding these mechanisms we are able to begin to enter the perceptual world of the animal. There are limitations imposed, for example, by the input systems. Thus the scientific paper 'What the frog's eye tells the frog's brain' shows how only large looming objects or small darting objects can be detected by the frog since that is all the cells in its eye can pick up. Both have clear survival value, the former warning of potentially harmful predators and the latter of possible prey in the form of small insects such as flies. Similarly we can attempt to enter the bat's world by investigating its echo-location system. The way in which the bat's brain codes the locations it picks up by reflected sounds has begun to be unravelled and the beautiful cortical maps of the various parts of the signal (distance, size, speed, and so on) are becoming clear. There is still, of course, great complexity in all of these animal systems, and especially in those of ourselves. Yet the correlation between brain and behaviour has grown enormously in the last few decades. It looks ever more likely that the evolution of Mind is identical to the evolution of brain.

Not that the purpose of this evolution is simply to have a brain of greater size. An elephant brain is three or four times bigger than our own, while that of a whale is over six times as heavy. The dolphin's brain is about the same size as ours. Yet none of those brains have the complexity possessed by our own, nor is their behaviour as complex. Even in our own species brain weight is not everything. The largest brain ever recorded was that of an idiot while, for example, the celebrated French writer Anatole France had a brain only a little over half that size.

There is another way to study the evolution of Mind. This is by observing and experimenting with infants and young children of different ages. Such study has led in general to a body of work indicating that as the infant and child grows physically there is a comparable development of its mental powers. This supports the identification of the brain with the mind. It also leads to the interesting question as to whether or not predetermined brain structures are present, independent of nurture, which allow the development of powerful behavioural patterns, especially language. The existence of such deep structures was strongly argued for a decade ago by the American linguist Noam Chomsky from his study of languages. These seemed to have common grammatical structures, leading Chomsky to posit prelearnt brain structures able to generate these common features across many human societies.

The study of learning in infants of say three, or seven, or ten months has been accomplished mainly by use of the technique of habituation. The infant sits in a comfortable high chair and looks at a picture projected in front of it on a screen. If several pictures, say of animals, have been viewed (with measurements taken of the length of time the infant looks at each) and the infant is then presented with a picture of one of these animals again, the time for looking may be shorter because the infant is assumed to have 'learnt' the picture. It may well look considerably longer at the picture of a completely new type of animal. If the animal pictures are suitably chosen the time the infant spends looking at similar or very different ones can be used to discover if it can create categories of animals in its head. Thus if it had looked at animals with a range of leg lengths it might not look for very long at an animal with an average of the leg lengths it had already experienced, whereas if it is shown a very long-legged (or short-legged) animal it may study it for considerably longer than usual.

In this way infants at three months of age have been found to be able to build up a repertoire of parts of such animals; at seven months a repertoire of single animals; at ten months the infant has apparently remembered a whole group of animals. A similar development of mental powers occurs for the ability to track objects going behind obstacles, so that the concept of 'object persistence' evolves over the same period. Face recognition seems to be achieved even at one month. Syllables appear to be recognizable over a broad range in the first few months, but such abilities seem to reduce and only months later does the power return, though now for those syllables heard naturally in the environment.

The evolution of Mind, either over the ages following the tree of evolution, or for the budding infant and child, does appear to be closely correlated with the evolution of brain. Yet it appears difficult to explain the powers of a Mozart, Shakespeare or Einstein in terms of the 'two fistfuls of porridge' of which the brain appears to be composed. It is only in the past century that the subtlety of behaviour has been matched by an appreciation of the corresponding intricacy of brain composition.

The brain

THE HUMAN BRAIN is composed of clumps of 'grey matter' joined together by white fibres. The clumps themselves are aggregates of nerve-cell bodies (occurring as irregular lumps or as sheets), the white fibres being outgrowths from them. The lumps of grey matter are termed 'nuclei' and given various Latin names, while the most important sheet is the cortex (or rind) which forms two crinkled half-spheres covering the brain stem and the midbrain nuclei in the mammal. It is the cerebral cortex which appears to have grown most with evolution, although a general increase in brain size compared to body weight (called encephalization) has been seen. This occurred for mammals, as compared to reptiles, about one hundred and fifty million years ago, for primates with respect to other mammals about fifty-five million years ago, and for hominids over other primates about five million years ago. During evolution over the last two million years, man's brain has about doubled in size.

The three parts of the brain – the brain stem, the midbrain

and the cerebral cortices – expanded at different periods of human evolution, and so perform different functions. The brain stem is the most primitive. Besides being a route for sensory inputs from the body and for muscle-control signals going down from the head there are control centres in it for the sleeping–waking cycle, for breathing and for simple reflex acts, such as walking and staying upright.

The next in level of complexity is the set of midbrain nuclei. These seem to be of at least three sorts. First are the relay nuclei that carry input from the outside to the cerebral cortices for further analysis. The most important of these is called the thalamus, which forms two sets of nuclei (one for each hemisphere). Vision, sound and touch, for example, are all transmitted for further cortical investigation through various thalamic nuclei from the primary receptors (eyes, ears, skin). The thalamus is shown, as are other parts of the brain, in Figure 1, opposite.

The second class of clumps of grey matter in the midbrain are those involved with giving significance to the various inputs. The most important of these – located in and around what is called the hypothalamus – was discovered in 1954 to be effectively a 'pleasure or pain centre' in the brain. An electrode had been inserted into a particular region of a rat's brain in such a way that the rat could press a pedal to stimulate itself by giving a small jolt of electricity to the electrode. It was observed that the rat would continually perform such self-stimulation and would even do so in spite of having to cross a relatively highly charged electrified grid. Various parts of the hypothalamus were discovered to be similar pleasure regions, others to be centres of pain as seen when the animal tried to escape from the region of the cage in which it could give itself stimulation if it were allowed to do so. This experiment was a turning point in the analysis of the relationship of brain to behaviour. Until then, the study of such a relationship was dominated by reducing the behaviour to a set of rules governing the way the conditions of the experiment determined the stimulus–response relationship. These theories had nothing to say about the inner mental world of the individual, which was banned from such discussion as being beyond the scientific pale.

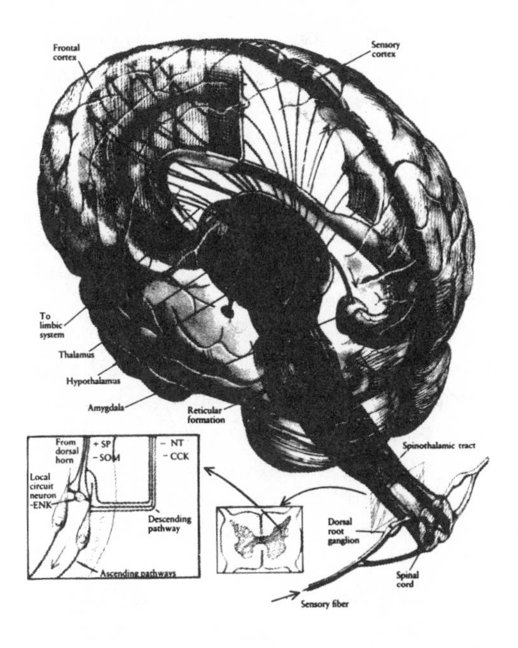

Figure 1: A schematic of the brain (from *Brain, Mind and Behaviour* by F. Bloom & A. Lazerson, W. H. Freeman, 1985)

But the results of the self-stimulation experiments were easiest to explain in terms of simply obtaining a pleasurable experience. Pleasure and pain once again became accepted as legitimate expressions. This was strengthened by reports from people who had been stimulated in similar brain regions. The experience, they said, was that of pleasure.

Besides these raw centres of pain and pleasure are other nuclei which seem able to store the significance of these pleasurable or painful experiences. One, termed the hippocampus (the Latin for 'sea horse', since this is what its shape resembles), is crucial for the long-term storage of experience. Without it one appears unable to learn anything new – words, names of people one meets, how to get back to one's house if one has moved. Even the ability to remember the beginning of the sentence one has started has gone. Another, the amygdala, appears to be involved in storing the emotional significance of some of the experiences of the past. Loss of the amygdala leads to difficulty in relating to other people. In laboratory animals, damage to the amygdala can attenuate emotional reactions to stimuli. Thus a rat with an amygdala lesion may continue drinking water made to taste unpleasant by adulteration with quinine. The nuclei we have mentioned so far – the hypothalamus, amygdala, hippocampus – constitute the limbic system (limbus denoting fringe or edge) since they are mainly on the edge of the cortex. The limbic system is clearly the repository of emotional content in the brain.

A third group of grey-matter clumps are called the basal ganglia. Their importance is in motor control and they appear to store programmes for movement. They are implicated in movement disorders associated with Parkinson's disease and Huntingdon's chorea. They are also importantly related to what is sometimes called the hindbrain, but more technically is the cerebellum. This is an elegant structure at the back of the brain which seems to be necessary for the fine control of movement.

Finally we come to the cerebral cortices – the two surfaces or mantles forming a covering to the brain and of prime importance in cognition. They have a layer of grey matter (comprising nerve-cell bodies) on the upper surface, with white string-like material (nerve-cell output wires, or axons) running underneath between different parts of the surface. The cortices are highly folded so as to increase the total cortical surface area that can be contained inside the skull. This is one of the ways in which the human brain has become more highly developed than that of the dolphin. The cortices can be roughly divided into three classes. First are the primary sensory regions for vision, audition, smell, touch and motor control. It is to these cerebral regions that various inputs (except for smell) arrive from outside, after being relayed through the thalamus. These inputs, after some preliminary analysis, are then further analysed in what are called the 'associative areas of the cortex'. These surround the primary regions. Thirdly, the highest level of cortex is in the frontal lobes which can be called the 'tertiary cortical region'. It is here that planning seems to take place.

The various areas of cortex – primary, associative and tertiary – have slightly different structures, even though all are built on roughly the same plan of six layers of nerve cells, with the cells grouped together in columns of about a thousand cells. There are well over a million such columns in the cortex. In spite of their rather similar appearance, the various cortical areas perform considerably different functions. This has been discovered by the increasing number of head wounds, as we remarked earlier. Damage to primary regions causes loss of sensation in the corresponding modality. The loss of the primary visual region in what is called the occipital lobe at the back of the head causes blindness. Loss of cortical cells in the associative cortex just in front of the primary visual region, in what is called the parietal lobe, can cause the strange phenomenon of 'neglect'. This means that a person may not be able to notice the left side of what would be their normal field of view. It may even be so extreme as to cause the person to deny that the left side of their body is their own, and they may attack their left arm or leg, claiming it to be an interloper in their bed. Loss of parts of the temporal lobe, at the sides of the cortices, causes loss of memory of particular episodes in one's past. Finally, loss of the frontal lobe can lead to a personality change and an inability to deal with social situations in a reasonable manner. A famous case is that of one Phineas Gage, who lost a considerable proportion of his frontal lobes when an iron rod destroyed them while he was working on the railway in the USA in the middle of the last century. He was reported as having walked, perfectly conscious, several miles to the doctor but suffered a considerable change of personality. He became quarrelsome and as a result died in a bar-room brawl some years later.

The brain is obviously the organ of the Mind. We will investigate in the next chapter how it has so far proved possible to explain its subtlety of action. Great strides are being made in this research; the 'mechanical Mind' is on its way.

Chapter Nine

BRAIN PROGRAMMES

The atoms of the brain

THE SCIENTIFIC revolution can be seen to be composed of two parts. The first of these, the 'material' revolution, has taken us in about three centuries down to the intricacies of quarks and gluons. On the way, the reductionist thesis has been amply verified, with ever greater precision. The second is the mental revolution, which has only been under way as effectively for just over a century. That may be one reason why, as we have already noted, there is such a discrepancy between our deep knowledge of the material world and our seemingly almost helpless state about our own selves. To put the powers of science into the hands of adults – especially politicians and the military – who are, in effect, still infants with little self-knowledge, is a step of great danger. No wonder that there are those in modern society who would wish to turn the clock back and return to the age of ignorance. Yet there are too many people on earth, too many mouths to feed and too high expectations of standards of living to stop the scientific revolution now.

The other reason why the mental revolution has taken such a long time in coming to fruition is that the understanding of Mind does not seem as simple a problem to tackle as does, say, water. The reductionist approach is clearly ideal for the material world. The gross properties of Matter, described by means of 'extensive' variables such as volume, temperature and pressure, can be explained by taking aggregates of atoms and averaging them. The resulting science of statistical mechanics allowed scientists to bridge the gap between the atomic and macroscopic levels. At the same time gross pieces of Matter hurling around – be they arrows, bullets or planets – were found to be governed quite accurately by the famous three laws of motion discovered by Newton in the seventeenth century. Quantum mechanics and relativity replaced these theories by even more precise ones at the beginning of this century, as we discussed earlier.

We would hope that the mental revolution could develop as rapidly by analogy with the material one. But to do this we need knowledge of analogues of temperature and pressure in the brain. Beyond that, where are the atoms of the brain, and what are the statistical arguments to allow us to explain global brain activity once the properties of those atoms are known? The answers to some of these questions are known, but the argument by analogy has not yet developed enough to allow as effective an understanding as in the material case.

To begin with, it can be rightly claimed that the atoms of the brain are the nerve cells or neurons, which have already been mentioned. Each of these is very small, usually no larger than a hundredth of a millimetre in diameter, and there are about ten billion of them in the human brain. All brain activity requires energy and an important percentage of the food we eat is used to keep the sodium/potassium pumps in our nerve cells functioning efficiently, in generating and propagating electrical impulses.

Nerve cells signal to each other in terms of a very brief wave of positive electricity, which flows steadily along the axon, like the burning of a fuse. A cell sends out such a signal if it has received electrical activity from other cells which exceeds a critical threshold amount. Thus the nerve cell acts as a tiny decision unit, responding if the electrical activity is above a critical level but staying silent otherwise. Each of these units is processing its inputs at about the same time as all the other units, so they achieve 'parallel computation'. This is now being realized as far more efficient than the serial method (in which only one unit would be operating at each time step) used in most modern computing machines. A lesson is being taken from the brain, and 'parallel computers' are being developed in which hundreds of processors work on a problem in parallel. However, it would seem that the brain is still the ultimate parallel computer, with its tens of billions of primitive parallel processors (the nerve cells). This is not quite the end of the story, because when the nerve impulse arrives at the axonal ending (called the 'synapse' from the Greek for 'to clasp', since axonal endings seem to clasp the adjacent cell wall) it causes a chemical transmitter to be released into the narrow gap (the synaptic cleft) between the axon and the next cell. It is this chemical which causes the change in electrical activity of the next cell. This alteration might be a positive change and so its action is called excitatory, or it might reduce the next cell's electrical activity as the process of inhibition. The synapse is also an important place where chemicals, especially drugs, can influence brain processing so as to give rise to hallucinations, depression or other changes in mood.

To summarize. The fundamental atoms of the brain are the nerve cells. They send out identical nerve impulses to each other along their axons if they are excited strongly enough (above a critical threshold) by other cells. There are two classes of nerve cell; excitatory and inhibitory. At nerve endings electrical activity is turned into chemical signalling (so explaining the vulnerability of behaviour to the actions of drugs of all sorts).

There appear to be about a dozen or so different types of nerve cell in the brain. It would seem, on the face of it, that the statistical (or averaging) approach to analyse the behaviour of a large number of such cells might lead to a useful description of brain activity, and begin to explain behaviour. However, the brain seems to be made up of connected groups of nuclei, as we described at the end of the last chapter. There is more subtlety in the brain than can be described just by taking an ever-increasing number of connected nerve cells and analysing them to determine their averaged behaviour. This subtlety is surely related to the complexity of behaviour itself, which is made up of components – emotional, cognitive, intentional, conscious, unconscious, pre-attentive and so on. There are no obvious global variables for such a complex system of a form similar to pressure and temperature in Matter. It seems necessary to decompose the brain into its modules or parts, where the 'global' variables for those components may become more apparent.

Preprocessing

THE INITIAL stages of processing in vision and audition are by means of out-stations of the brain which play a very important role. They act as filters for the brain itself, so that 'what the frog's eye tells the frog's brain' is quite limited, as we remarked in the previous chapter. Even the mammalian retina limits what the brain can work on. However, as the evolutionary tree is ascended there seems less and less that is filtered out at the retinal level in an animal, more work being left for the cortex to perform (which we noted earlier had expanded as a consequence of increasing demands). It is necessary for us to know what preprocessing is being done. Otherwise we cannot properly specify the problem posed to the rest of the brain as it tries to make sense of the incoming message, then store it, label its significance, and make decisions regarding it, to allow appropriate goals to be reached. It is also valuable for us to study what happens in particular preprocessing organs, such as the retina, cochlea and in olfaction. The first of these has been the object of exhaustive analysis since it has rightly been called an 'approachable part of the brain'. The hints which the retina (or other similar units) give us as to what sorts of processing may be achievable in other areas of the brain are of value. I will concentrate on the retina.

A visual scene is often made up of areas of the same intensity of illumination bounded by edges and corners. The regions of constant illumination or colour contain little information, so need not be transmitted to the brain for further processing. They are redundant. Such redundancy can be removed by sending on from the retina only the difference between the size of the visual intensity at a certain point and the average intensity over a surrounding region. If the intensity of the surround is the same as that at the point in question then no signal is sent to the brain; redundancy has been removed. This method has been used in sending television signals in a much more compact form over transmission lines. Our own retina appears to use that principle most effectively. It is achieved by having a layer of cells (called the outer plexiform layer) which are highly interconnected, one cell to its neighbours, by means of even closer contacts than occur usually between nerve cells in the brain. These connections send an electrical signal from one cell to the next by means of very narrow junctions, which join the dendrites (outgrowths from cell bodies) of the cells together. This is why they are called dendro-dendritic synapses. As a result, the outer plexiform layer is like a net of electrical resistors, and this has been modelled very effectively as such by electrical engineers, using large-scale integration methods to produce a very compact device – the so-called 'silicon retina'. A silicon ear has also been made, using similar biologically based design principles.

An interesting feature of the transformation the retina makes on an input is to enhance edges. As an edge is approached from the darker side the averaged value of light intensity is increasingly positive and this is subtracted from the receptor cell output at the point, which is zero; a negative response results. On the lighter side, across the edge, the reverse change happens until far enough away from the edge there is no difference between receptor cell output and nearby illumination. There is then no response. This means the edge appears exaggerated with its dark side made darker, its brighter side even brighter. This edge-enhancement effect has also been discovered in the invertebrate retina, which uses a rather similar design principle, now termed 'lateral inhibition' because cells inhibit the response of those on either side. The invertebrate retina is composed of lots of little 'eyes' called ommatidia, all receiving light and processing it in parallel. Light falling on an ommatidium in the retina, say of a crab, causes a reduction of the signal being sent further on from nearby ommatidia by connections from the first unit trying to inhibit laterally the output of its neighbours. This can be observed quite clearly if the output of a single receiving unit is monitored while a thin pencil of light is shone on it. A similar beam shone on a nearby ommatidium causes the response of the first unit to be reduced; a thin pencil of light shone on a further unit laterally inhibits the second unit, and so releases or 'disinhibits' the first. The principle of lateral inhibition and the process of disinhibition are ones which are proving valuable in explaining processing occurring in higher brain regions.

There is further processing performed in the retina of vertebrates after light has passed through the outer plexiform layer described earlier. In lower animals like rabbits some cells have been found which are responsive to the motion of objects in a particular direction, so are called 'motion detectors'. Our own retina does not seem to contain such motion sensitive cells but delays the analysis of motion until the information reaches the cortex of the cerebral hemisphere. This makes our human system more flexible because it can analyse many more possible sorts of motion than the rather more stereotyped motion detectors in the rabbit.

Object recognition

THROUGHOUT VISUAL processing a very important concept for each nerve cell is that of its 'receptive field'. Imagine a screen say a metre or so in front of you. Let a spot of light shine on it in a particular place. Then that region of the screen which causes a cell in your retina or in your visual cortex to respond is called the 'receptive field' of that cell. The output cells of the retina have circular receptive fields, which usually subtend less than one degree across at the eye. This is little changed when the input reaches the visual relay part of the thalamus. This is the relay centre at the top of the brain stem en route to the cerebral hemispheres. Nor is the visual signal much changed by the time it arrives at the input cells in the visual cortex at the back of one's head.

There are other input cells in this visual region which act as slit detectors. It is as if the visual picture you see in front of you has been split up into all its edges and contours, and only they are being noticed. This orientation sensitivity changes gradually as one proceeds across the surface of the brain, so that after travelling about a millimetre the orientation has rotated through a total of about three hundred and sixty degrees. All these cells have roughly the same receptive field position.

There is an added complexity in humans in the form of colour sensitivity. It appears that every millimetre or so there is a 'blob' of cells. These cells are sensitive to coloured input at that position on the illuminated screen but are not sensitive to oriented inputs. This means that colour and intensity have already been segregated into two streams of information. There is a further separation in that the transient and sustained outputs from retinal cells are handled by separate cortical layers and are used in segregated streams later to give separate motion and intensity channels of information.

To summarize. In the early stage of processing in the visual cortex there is segregation of input into different streams carrying information about the colour, the motion and the intensity in the picture being viewed. The intensity is described by a set of slit detectors, splitting the visual input into a set of edges. Each of these streams of information arises from a different small region in the input. This is an extremely flexible formulation of the visual data on which to work, but poses some severe problems.

First of all, various aspects (edges, colours, motions) of the objects of the visual field have been split up into those features at a given point. The image has been divided up for further processing but it must also be put back together. Otherwise, objects in the visual scene could not be recognized. My dog is to me, when I look at her, made up of separate little spots each with its own features. How do I recognize her as the whole dog? How do the colours and intensities fit together so well? If she is moving how does the motion information and the colour information let me recognize that it is my dog which is running across the floor?

Several steps towards this are taken in succeeding stages of the visual cortex. By combining inputs from different sets of cells with different receptive field positions at the next stages of processing (in cortical areas) it has been observed that the receptive fields of cells double successively, to end up finally as the whole of visual space. In the process, cells are able to recognize larger and larger parts of objects, until whole objects are responded to in visual cortex.

The simple additive recombination of what had been originally split does not solve all problems, however. Motion, colour and edge information have been separated out. They have to be recombined so that when I look at her the dog I see is golden brown and has not accidentally acquired the colour of the grass on which she is lying. Indeed brain damage can cause misalignment of colour and shape so that the red of a dress might appear to extend over the face of the wearer. The problems that are thus faced are of image segmentation (separating features of an image from its background) and then the binding together of the various featured parts (colour, edges, motion) into which the image had been split.

There have been recent discoveries in visual cortex that hint at a solution to the segmentation problem. Inputs from the edges of an object have been found to cause synchronized activity among the cells in the visual cortex acting as feature detectors for them. This activity may be oscillations at about 30–50 cycles a second. The co-operative activity of these cells triggered by a given object has been found to be different from that for another object in the scene. These two sets of neurons, one for each object, are thus responding in a different fashion to the different objects. The problem of fusing together the different features of a given object has not yet presented such a clear solution. Various ways of labelling the responses of the ensembles of cells coding for the differing features have been suggested. However, there is one aspect of visual processing of which it does seem important to take account. Attention plays a crucial role, seeming to act in a manner contrary to what we would expect.

I mentioned earlier that one of the important steps taken recently to speed up large-scale computing has been the development of parallel computing. Several microprocessors are connected together so that job-sharing occurs and so that numerous elementary tasks can be performed in parallel, one on each of the microprocessors. As I noted then, the ultimate parallel computer is a network composed of neurons. Each of these latter appears to perform a decidedly simple operation on its inputs. The neuron decides if the sum of its inputs is larger than a preset threshold, in which case it sends out a brief signal: 'I am on.' Otherwise it is silent. Artificial neural networks made up of such simple neurons would seem to be the ultimate parallel computers. The neurons all act in parallel, and do not have to wait their turn. This is one of the reasons why the use of artificial neural networks has burgeoned in the last decade. They avoid the 'one-at-a-time' serial processing which slows down standard computing. Artificial neural nets are therefore composed of many neurons acting in parallel. Such networks of ever-increasing numbers of neurons are being used to solve interesting tasks in factory and robot control, prediction, vision and speech recognition. We must remember that the artificial neuron is only a caricature of a living nerve cell, so that in calling a neural network an 'artificial brain' one should emphasize the 'artificial' rather than 'brain'.

As the number of input lines to a living or artificial neural net increases due to increasingly complex inputs (visual, auditory or olfactory) the problem of, say, recognizing the input as one of a class given beforehand becomes more difficult. This is particularly true if the connection weights between the neurons have to be modified in some way in order to solve the task. These connection weights are numbers which indicate how strongly the output of one neuron influences another neuron further on. Because there are more and more weights to change, the problem could well become insoluble. It seems that we could help resolve the difficulty by having parallel processing being effected at various scales. But ultimately, and in a way not yet properly understood, 'attention' is used by ourselves to filter out unlikely candidates and cut down the difficulty of the problem. Only a small part of the total visual field is inspected. For example, in searching for a cup only the inputs in a region round that cup will ultimately be analysed. This is achieved by directing attention to the cup, thereby reducing the receptive fields for the relevant visual nerve cells in early cortical visual processing areas. How we move our attention is very interesting, especially because adapting the technique to the artificial neural network situation might have important implications.

There are various clues as to the nature of this attentional process in humans. For example, if a picture on a screen, say of a P, is searched for among a set of distractors such as Ts, the target P 'pops out' easily, in a time independent of the number of Ts in the scene being looked at. If the distractors are Rs the search for the P takes longer because Rs and Ps possess similar features, as can easily be seen. In fact the search for the P among Rs takes a time proportional to the number of distractor Rs. This 'linear' search for a target possessing a conjunction of the right features is one which indicates that there is some sort of competition going on between the various features vying for recognition by higher-level control systems, which must carry a representation of the target.

At the same time, studies of brain-damaged people indicate that there are at least three aspects of attention, controlled by different regions of the brain. First there is the action of disengagement from present fixation of attention, apparently controlled by regions in the parietal lobe just in front of the visual cortex. In particular the information stored there is about where objects are, not about their details. Then there is the process of movement of attention, which requires integrity of a midbrain region near the thalamus called the superior colliculus. Finally the action of attending to a new object must occur, for which some sort of competition between possible candidates is to be expected. Recent observations on increased blood flow in human volunteers when they were searching for visual targets of letters among other letter distractors have shown that this re-engage process is controlled by increased activity in a particular part of the thalamus. Once attended to, visual input then enters further neural networks in the temporal lobe (at the side of the head) on the way to the memory regions in or near the hippocampus, mentioned earlier as crucial to memory. If the object has been experienced earlier, then conscious recognition can occur and further objects or goals turned to. On the other hand, if the object is unfamiliar it may be further inspected to discern its nature. The manner of the coding of objects in this region is still somewhat uncertain, although faces seem to be recognized by their edges while other objects are recognized by surface and volume features. It is still not known whether objects are encoded by means of the hierarchy of their parts (my dog as a set of dog legs, dog head, dog hair, etc.) or otherwise.

Memory

IT HAS ALREADY been noted that memory requires a particular organ, the hippocampus, to be effective but that memory storage may occur in other nearby regions of the brain. Yet memory is an amalgam of various sorts of remembrances. Patients with total amnesia for, say, new faces, are able to give away the fact that they can recall a recently experienced person not by conscious recognition but by a change of their skin resistance on the palms of their hands. As part of tests to discover what abilities they still possess, people with no hippocampus have, surprisingly, still been found able to learn a skill such as upside-down or mirror writing, although how this occurs is unknown. They cannot consciously recall having been trained to perform in such a manner, yet the skill at such writing has not been lost from the time when they had been trained to do it. There would seem therefore to be a range of sorts of memories: skill memory, recognition memory, recall memory, each with its own modes of operation, but only the latter involving awareness. Moreover there is apparently a further distinction between memory for specific episodes, such as where one was when one heard of the assassination of President John F. Kennedy, and memory for words or similar concepts. Episodic memory involves, it would seem, conscious recall while semantic memory for words does not appear to need awareness. I am not aware of each word I write or say until the instant of its production. People who have had signals of a word flashed briefly on a screen in front of them, and then a white flash to mask out the word, are not consciously aware of the word, since it has apparently been blanked out before it can erupt into consciousness. However the person is still influenced by the word, in that if asked to choose between their word and another similar one, they will choose the former. It would seem that visual word-recognition units may be set up to function in a pre-attentive manner in much the same way that oriented slit detectors process early visual information without awareness.

To summarize. Memory in ourselves seems to occur in at least two forms. One needs the hippocampus to function, and involves conscious processing. It would seem that this awareness would have to be present during the initial experience in order to recall it. The other is extra-hippocampal and unconscious, and includes motor skills, word recognition and possibly other non-attentive aspects of learned responses, such as the unconscious face recognition mentioned earlier.

The way in which connection strengths between living nerve cells could change during experience (so storing a memory of it) was discussed in the late 1940s. It was proposed that these connection weights could change by amounts dependent on the activity coming in on the input line to a neuron, and on the concomitant output activity of the target neuron. If both input and target cells are active, this correlated activity should be strengthened by increasing the connection strength. If the input cell is inactive and the target cell is active, that lack of correlation should be noted by reducing the connection weight. If there is competition between neurons, so that only the winner and its neighbours learn, then a topographic representation of inputs can arise, as is found to occur in the visual cortex. There the visual image is found to be mapped (with some distortion) on to the surface of the brain in a topographic manner. This approach has been used in its application to artificial neural networks and has led to a wide range of industrial applications.

The sort of learning originally suggested involved no teacher to say exactly what the response of each neuron should be to a given input, so it has been termed 'unsupervised'. An alternative mode of teaching is to reward the neurons when they fire in a manner which leads to the correct final output and punish them if they produce an undesired response. This so-called 'reinforcement learning' increases weights if they are rewarded and decreases them if they are punished. It has been found to lead to useful artificial neural network learning systems for training to solve certain tasks, such as robot or industrial control. It may also arise in animals, where rewards can be used to train them in all sorts of ways, as performing animals demonstrate in a circus.

The final approach to learning is that which uses complete supervision. A 'teacher' says exactly what outputs should arise from each of a given set of inputs to a neural network. This supervised approach has become very popular in the artificial neural network field, especially in the last decade (although is less appropriate to explaining learning in animals). This is because a solution had been arrived at of how to train 'hidden' units. These were so called because their desired output was not given explicitly: they only sent output to other neurons in the net and not to the teacher. It was discovered how to assign credit or blame to each of the neurons in the net for the output, be they 'hidden' neurons or not. Numerous supervised learning methods in artificial neural network research and development are currently enjoying great vogue throughout the world. These methods are being used to solve important industrial and business problems. An area such as the prediction of future trends in the stock market already has neural networks which are trained to produce better predictions of exchange rates than human dealers. One can envisage a time in the future when the financial markets are controlled by competing neural networks, each of them being constantly improved by their research teams.

The process of 'reverse engineering' takes properties of the brain and applies them to solve industrial problems. This has been used extensively in a wide range of tasks. The opposite process has also been used to apply understanding gained from artificial neural networks to allow a better understanding of memory storage possibilities in animals, both in hippocampus and other regions of the brain (cortex, basal ganglia, cerebellum, etc.). For example, it can be calculated that roughly four to six weeks of storage capacity may reside in the human hippocampus. That calculation is based on very simple model neurons so as to reduce the complexity of the neural network and make it easier to analyse. But living neurons are not simple, so the quoted storage capacity of the living hippocampus may be an understatement. However, the principle of using simple neural network models to try to explain living brain activity may help if we want to discover the global processing methods. For example, increasing complexity can be used. Too great a complexity, however, may be inappropriate to begin with; we may never see the wood for the trees.

Goals and plans

DRIVES RESULT from a state of the body, although for humans (and possibly primates) there are drives attached to achievement and to being well liked. These latter drives can be said to be psychological, as is that causing a condition like anorexia nervosa. The resulting goals are those objects which will satisfy the drives, such as a good steak and chips for hunger.

The drives which are aroused are due to various chemicals causing activation of limbic areas. They are not activities which will be reduced without satisfying the goal; hunger is difficult to forget until one has eaten enough. The goals are, however, not only specific objects of desire but are also stored representations.

All these processes involve the use of stored representations of rewards (for example, food) which are associated with and excited by the corresponding drives. Sets of motor responses must also be related to the rewards in order to indicate how the latter could be obtained. The associations between these various stored representations (drive plus goal plus motor response) are probably developed by learning as a result of our very many life experiences. Drives would be expected to activate a number of possible rewards, but only a few – perhaps only one – might be chosen as the most appropriate in the context of the input stimuli of the given environment.

This process of decision making would seem to involve conscious awareness of the decision to take. Some people make lists of possible steps that could be taken and grade them according to the pleasantness of their results. Planning performed in the mind is probably accomplished with the use of limbic structures for evaluation and by calling on stored experiences, while the detailed evaluation of consequences could well occur in parts of the frontal lobes (sadly lost to Phineas Gage, mentioned earlier, along with his powers of forward planning). All these processes seem to involve conscious awareness at some point or points along the way. I discussed discovery of pleasure/pain and memory centres in the limbic system earlier. However, there seems to be no comparable consciousness centre. In the case of split-brain people, they seem to have a separate consciousness for each of the now-separated hemispheres. Does consciousness arise from whole-brain activity, so that it is a distributed property: or is it impossible ever to explain in physical terms?

Chapter Ten

CONSTRUCTING THE MIND

Mind versus Matter

THE SIMPLE BUT strong response 'What is Matter? Never Mind. What is Mind? Never Matter!' is the unthinking one of many in society. Francis Crick, one of the discoverers of the structure of DNA, recounted how he once got into conversation with a lady sitting next to him in an aeroplane. To her question about what he did, he responded that he was trying to understand the nature of the Mind. The basic question, he tried to explain, was how was she able to experience the world around her in the way she did. 'Oh, that's easy,' she is supposed to have said. 'There is something like a television set in my brain with a picture of the outside world on it.' 'But who is looking at that?' Crick asked her. At that point she realized why the problem was not so easy after all.

Indeed the problem is very difficult. Philosophers over the millennia have written many learned tomes on it, which seem to cover all possible answers. These range from the idealist extreme, where the whole universe is composed ultimately of 'Mind stuff', to the other where only Matter exists and the Mind is a mere epiphenomenon, albeit a very subtle one, of the activity of the brain. The success of the scientific revolution indicates that it is very unlikely that the Matter being probed to ever greater precision and understanding in our quest for the theory of everything is a mere figment of the imagination. It has become ever harder to deny that there really is 'a physical reality out there' independent of ourselves. Scientific progress has been made by assuming that this is so, and we will go along with it. But it is also clear, from the effects of drugs, head wounds, meditation and similar phenomena, that the material of which the brain is made determines to a very important extent the mental experiences that can be had by the Mind accompanying that brain.

There are two possible conclusions one can draw from such dependence of Mind on Matter. One is that, in spite of the hint that seems to be given that Mind is really some aspect of Matter, this is not actually the case. Mind, according to this approach, is an independent entity which is connected to Matter in an as yet unknown manner. Moreover the mode of characterization of Mind is also yet to be clarified. This conclusion is usually termed 'the dualistic approach', and has had some quite recent supporters. These include one of the doyens of brain research, Sir John Eccles, who wrote in 1977: 'Briefly, the hypothesis is that the self-conscious mind is an independent entity that is actively engaged in reading out from the multitude of active centres in the modules of the liaison areas of the dominant cerebral hemisphere.' Such dualism essentially posits that science can only go so far in understanding the Mind. Thus Eccles states: 'But the question: where is the self-conscious Mind located? is unanswerable in principle.'

The alternative conclusion to be drawn from the powerful effects of Matter on Mind is that Mind is caused totally by Matter. The construction of Mind-like states from what is apparently unthinking, unfeeling chunks of Matter is a challenging problem whose solution must be attempted. This appears now to be happening in brain research, with increasing numbers of conferences, journals and books devoted to the issue. It signals the defeat of the strict behaviourist school of psychology, which claimed to be able to explain all behaviour solely in terms of stimulus–response reflex arcs and their reinforcement. The use of internal representations of past experiences was regarded as unnecessary, and even misleading, by the ardent followers of the behaviouristic school. Yet the denial of thinking as an aid to action left the approach too weak to explain complex animal acts which clearly depended on internal states of the animal. For example, there are a number of experiments showing that animals can learn a good deal about their environment without any external reward being present. Moreover, a rat in an unbaited maze runs less and less the more it learns because its curiosity is diminished; by strict behaviourism it should run more and more in a search for reward. A cognitive approach to this and many other animal-learning phenomena appears to be needed.

I will follow the second of the two approaches to the Mind–body problem and assume that Mind emerges from suitable activity of Matter. This latter itself is the ground of all being. Such an approach is economical, in that there are not two different forms of being, that is, Matter and Mind, and with them the two apparently insoluble problems of (a) the nature of Mind (noted in the earlier quote from Eccles) and of (b) the interaction between Matter and Mind. Instead we are left with the hopefully more tractable problem of constructing Mind from Matter. However there are several claims that the problem is still impossible.

The thrust of the complexity argument against constructing a physical model of Mind is that the problem will always be too difficult to solve. The brain is composed of billions of neurons, all of them extremely complicated with interconnections and functions which, so it is claimed, we will never be able to probe effectively. Even if we could comprehend the mode of action of, say, the early visual cortex in sight or of the olfactory cortex in smell, these are illusory gains since the whole of the brain, with all the areas working in combination, is infinitely more complex. This difficulty is one which neuroscientists have appreciated. They responded to it in the past by trying to probe ever more deeply and at ever smaller distances, following the reductionist path. They have begun to learn a great deal about the way various chemicals are effective in causing changes in the activity of certain nerve cells. But neuroscience seems to be fragmenting into ever greater specializations. Even in neuroanatomy a colleague will say, 'Sorry, I don't know anything about organ A – I'm an organ B man myself.' This is due to the increasing depth of science – a frequently quoted description of the scientist is: 'Someone who knows more and more about less and less.' One can compare this with the time of Milton, when an educated person knew science as well as the arts. Such specialization is undoubtedly making the thrust of the complexity argument stronger.

It is indeed true that nerve cells are highly complicated devices, and that aggregates of them are even more so. The range of chemicals involved is quite bewildering. Yet there must be a set of overall programmes or modes of action of the whole brain in terms of which its action could be explained. That explanation should contain, as an essential feature, the emergence of consciousness.

I said earlier that there has been an upsurge of interest in consciousness, with a growing number of journals, conferences and books devoted to the topic. As I observed in the last paragraph there is increasing compartmentalization in neuroscience; the upsurge of interest in consciousness seems to be strongest among philosophers and psychologists. In spite of the perspicuity and insight brought about by these endeavours they do not seem concerned with the actual machinery of the brain. A recent philosophical book on consciousness notes: '. . . our anatomically noncommittal account of competition between coalitions of specialists . . .' in discussing which cortical activities (the specialists) guide conscious activity. Another recent book on consciousness has only a brief chapter on a possible neurobiological implementation of the 'global workspace approach' (in which processed inputs are broadcast around so that each is known about globally through the cortex) and then the brain is ignored in further discussion. As might be expected, neuroscientists are also beginning to contribute although they may be limited by their specialization. A recent attempt to explain Mind by drawing on Darwin's principles of natural selection has led, according to one neuroscientist, to something that did not behave like a machine but like a little animal, perhaps a sloth. Neuroscientists appear not to appreciate how complex real machine behaviour can be.

This neglect of the possible way the brain could act so as to be the mechanism of Mind leads to the weak chink in the materialists' armour. For no engineering blueprint has yet been provided to persuade one that there may be a viable mechanism for creating consciousness from brain activity. There are few, if any, predictions as to crucial experiments which could destroy any such materialist theory. The enterprise appears dangerously vulnerable to the dualist's or idealist's response – 'It is not convincing.' This was effectively the critical response given to Rodney Cotterill's recent book, *No Ghost in the Machine*. In a carefully researched thesis, the author stated: 'I have come down rather strongly on the side of determinism.' Determinism is the principle that human mental experience is completely determined by physical brain activity, and there is no free will. Although the author's account of the power of Matter over Mind was heavily supportive of his claim, and of his book's title, there was no hint as to how consciousness could actually arise from the undoubtedly delicate and subtle activity of the brain. The loophole was not closed.

My emphasis in this chapter is in trying to avoid that fate and to attempt to close the loophole. This requires talking about the 'tissues' of the brain, which is not popular with the so-called 'functionalists'. They regard the function of the brain as paramount and its actual machinery as unimportant. There is strength in that position but also the weakness, noted above, that it is not convincing unless a blueprint is ultimately forthcoming. To put forward a possible detailed mechanism for consciousness I will attempt to use the general information-flow characteristics observed in the brain and in mental states as well as possible clues that may be given by the global wiring diagram of the brain. One might hope that through a sympathetic balance between function and implementation the hints to such a blueprint for a 'Mind machine' might become apparent. It might be possible in such an interdisciplinary manner to bring together the insights of the philosophers and psychologists with those of the neuroanatomists and neurophysiologists.

Are thinking machines logically possible?

JUST OVER thirty years ago it was claimed in a highly stimulating article that humans could not be machines. This was based on fundamental work done in the 1930s relating to the question of the existence of undecidable statements. These are deductive statements, like theorems in mathematics, whose truth or falsehood can never be decided. It had been shown then that such statements can be constructed in any language which contains sufficient grammatical structure. More recent work pointed out that in spite of the undecidability of such statements we, as humans, can see that such statements are true. It is as if we can use a sort of language superior to the one for constructing these undecidable statements. This would mean that any machine which was constructed to argue, using one or other of these languages, would be limited in comparison to ourselves. Since any machine is constructed to obey some language or other (its underlying programme) then such machines could never argue – or think – as powerfully as we can. Therefore humans are not machines. QED.

This claim, supposedly based on logic, that humans cannot be machines, created a furore at the time and led to a host of counter-attacks in a range of learned journals. Many arguments against this thesis were presented, with titles such as 'Lucas's number is finally up' (John Lucas being the originator of the claim). However, in spite of their erudition, none of them seemed to hit the mark. They attacked all the seemingly weak points of the claim, but Lucas rebounded, thinking fast on his feet, and apparently remaining unbeaten and unbowed.

The problem he posed for a purely materialist approach to thought (sometimes called strong A(artificial) I(intelligence), by virtue of the fact that many members of the AI community believe that thought can be represented solely by chains of arguments) is as follows. Consider some language L, and the sentence 'This sentence is unprovable in L.' Is this sentence true or not? There are four possibilities: either the sentence is true (or false) and provable (or unprovable). There are no other possibilities. But it cannot be true and provable (since if it were true then it must be unprovable), nor can it be untrue and provable (for if it were provable then it would, *ipso facto*, be true, since this defines a proof, and therefore it is inconsistent with its assumed untruth). Finally, it cannot be untrue and unprovable (since if it is the latter it is true). Hence the sentence can only be true but unprovable. We can see it to be true, by the above argument, but it cannot be proven so using only the formal language.

This result clearly leads to severe limitations for the approach to thinking by strong AI. For example, it would not seem possible to build a machine able to solve worthwhile problems in mathematics. Mathematicians must resort to non-algorithmic or creative tricks to construct non-trivial mathematical structures and prove hard theorems. This has been resuscitated recently, with the claim that we must modify the laws of physics used to describe the brain to bring in quantum mechanical aspects, especially those of quantum gravity. But this appears to be highly unlikely because the energies involved to harness quantum gravity are many orders above those available in the brain (which only has electron volts) and have not yet even been reached by the latest gargantuan particle accelerators at CERN or in the USA. The non-algorithmic processing occurring continually in our brain should have a simpler explanation than having to be fired up to very close to the speed of light in a particle accelerator to obtain the energy needed. I will argue shortly that instead of modifying the laws of physics we should attempt to understand the details of the mechanism of consciousness and at the same time glean insight into unconscious processing of mental activity. Only then can we expect to come to terms with the creative process, be it in mathematics or the arts, in physics or literature, in biology or music.

Global clues

As I REMARKED earlier, neuroscience has made progress in analysing the brain by studying ever smaller portions of it. But that has not helped much in understanding how the brain works in a global manner. What is needed is a flow chart showing how inputs move from one region to another and finally result in motor actions. In between, all sorts of emotional assessment, memory reactivation, decision making, and so on, will have been traced. This has proved very difficult to achieve, in spite of the myriad of damaged brains producing defective processing which have been studied, and in spite of animal experiments which test in detail ideas on functions of particular brain regions. The defective brains are limiting in that they usually involve damage to several regions and, of course, not all brains are identical physically. The work on animals may be helpful in understanding the animals themselves, but they are not necessarily perfectly analagous to the human brain. If they were, it might well be the animals experimenting on us!

Important advances have occurred, however, over the previous few decades, and most particularly over the last one, in the development of sensitive techniques to measure activity throughout the whole brain during the various states of attention, inattentive wakefulness, sleep, etc. These techniques involve methods that do not require surgery, for example by implanting permanent electrodes through holes drilled in the skull (as occurs in the case of epileptics in order to discover which part of their brain is producing the epilepsy, so that it may be cut out later). Such electrodes are termed non-invasive and can be used on any human subject with little or no interference or discomfort. The techniques include measurement of the low levels of electric current circulating on the scalp (termed electro-encephalography, or EEG) and of the very low magnetic fields produced by brain activity (termed magneto-encephalography, or MEG). There are other approaches, for example monitoring blood flow in certain brain regions by the increased emission of particles from radioactive material ingested by the subject, and studying the effects of magnetic fields on the material of the brain. These latter techniques have been used in investigating the brain regions involved in attention which were mentioned under 'object recognition' in the previous chapter.

EEGs are measured by attaching small metal electrodes to the subject's scalp. The minute electric disturbances, which are then highly amplified, are a signature of underlying brain activity. These were first measured on the intact scalp in 1929, but disregarded until some years later. The interpretation of the wavering traces of activity from different regions is still something of an art, although it has proved of great importance in detecting focal centres of epileptic activity and of brain death. However even such expectedly clear features are debatable, since EEGs cannot measure deep brain activity where epileptic seizures may actually be explosively initiated before rumbling away on the brain's surface. Flat EEGs, usually taken as a sign of death, can occur if somebody has become very cold, is very badly poisoned, or in certain children: in none of these cases is the person actually dead, yet this is what the EEG would have led one to conclude.

Advances in EEG analysis are, however, steadily occurring. There are characteristic features of the level of synchronization and speed of variation of brain waves according to the state a person is in when the waves are being measured. These can roughly be summarized as four different rhythms: Alpha (8–13 cycles a second, occurring when the mind is at rest), Beta (greater than 13 cycles a second, and occurring when attentive), Theta (4–8 cycles a second, when drowsy) and Delta (below 4 cycles a second, occurring only in deep sleep). A fifth state, Gamma (25–60 cycles a second, and observed when a subject is awake), is of great interest in assessing whether or not a person is conscious under anaesthesia.

Consider the following reports of patients who were paralysed while under anaesthesia for surgery:

> The feeling of helplessness was terrifying, I tried to let the staff know I was conscious but I couldn't move even a finger nor eyelid. It was like being held in a vice and gradually I realized that I was in a situation from which there was no way out. I began to feel that breathing was impossible and I just resigned myself to dying.
>
> I came out of the anaesthetic and couldn't understand why I wasn't in the ward. I could see the surgeons at the end of the operating table and I thought 'Oh, my God, they're going to operate on me and I'm awake.' I tried to tell them but I couldn't speak – couldn't move – it was the worst experience of my life.

Thus the patient is conscious but cannot indicate that he is. Apparently, recall such as in these cases occurs in less than one tenth of a per cent of all patients given a general anaesthetic, but since there are about thirty million operations a year carried out in the USA alone, there could well be thousands of such cases. Recent research has shown that a key index to the loss or otherwise of consciousness could be the presence in the auditory cortex of a gamma type of response to a brief sound. This EEG response pattern is observed after averaging a number of presentations of auditory clicks to a normal human subject. The response occurs within a hundredth of a second after the patient has heard a click, and continues for about a tenth of a second afterwards. Under general anaesthesia this characteristic gamma oscillation should be absent. However, there is anecdotal evidence that when an anaesthetized patient has periods of spontaneous movement or moves in response to a spoken command there are always related gamma oscillations which can be observed in their EEG traces. This has led to the suggestion that as long as gamma oscillations are present then the patient is still, to some extent, conscious. Such gamma oscillations have also been observed in single neurons in the visual cortices of lightly anaesthetized cats when presented with illuminated regions. In fact, pairs of nerve cells have been found to oscillate very closely in parallel under such conditions, even when the cells are in opposite hemispheres. Similar oscillations have been found in midbrain nerve cells, especially in the relay station of the thalamus, mentioned in the previous chapter. These effects may be important for consciousness on a more general scale.

The other technique, of MEG, has produced exciting new findings which might begin to tie up results on cortical and midbrain (especially thalamic) activity. This is because magnetic fields can penetrate through the medium of the brain. With enough care taken over the experimental set-up, it is thereby possible to pin down the positions of those nerve cells whose activity (firing) causes a small magnetic field to be produced. The set-up itself requires several magnetic field detectors (up to thirty-seven in recent experiments) similar to the way that EEG measurements may use sixteen or so electrodes on the scalp. The magnetic field detectors are positioned round the subject's head and have to be specially designed to be sensitive enough to pick up the very low level fields emitted (several orders of magnitude below that of the earth). They use the superconducting material mentioned in an earlier chapter. At the same time, the tests have to be carried out in a very carefully designed, magnetically shielded room to cut out any effects from changes in the earth's field, passing motor cars, etc.

The most important finding, although still controversial, is that when attention occurs there appears to be a global sweep of gamma activity. This is started in the front part of the head and moves backwards quite rapidly. Moreover it would seem that an area of the midbrain, probably the thalamus, is the originator of this activity, since it precedes the cortical activity by three milliseconds. This is very close to the time taken by nerve impulses from thalamus to ascend to cortex. In addition, the relative strengths of the cortical and thalamic activities during this sweep seem highest for healthy youngsters of, say, twenty years of age, and decrease for older people. They are apparently even more reduced for patients with Alzheimer's disease.

The presence of this sweep, and the variation with age of its relative cortical to thalamic strength, explains a number of ideas and features of gross brain activity. In particular, various neuroscientists had suggested in the 1960s that the frontal lobes produced activity predicting the input entering the brain at primary sensory regions. This 'corollary discharge' theory is seen here as being manifest by the front-to-back gamma sweep. Predictions from the frontal lobe are fed back to where inputs are still being processed, and a comparison can then be made. There is much more experimentation to be performed before such an identification is validated, but at least the possibility is very suggestive. A further feature of interest is the increased dominance of midbrain over cortical activity as ageing occurs. This may possibly be related to the noticeable increase in inflexibility of human beings as they age. Instead of ascribing it only to a 'hardening of the arteries', the increasing age-related thalamic dominance over cortex (due to weaker cortical connection or for some other reason) may also play an important role.

The relational Mind

EVER SINCE I was quite a small boy I have puzzled over the nature of Mind. I can remember walking over the Lancashire moors with my father when I was twelve or thirteen years old listening to his account of philosophers' theories of Mind, and his response to them. I went on to train and research as a theoretical physicist, since it seemed to me that I should first learn all that I could about the material world. That I did, and I have tried to recount some of that understanding in Part Two of this book. However, I have always retained the desire to understand the nature of Mind. I reached as far as I felt I could go in analysing superstrings some few years ago. I also realized at about that time that there is probably no theory of everything, and that high-energy physics would go on for ever probing the Universe. On the other hand, a revitalization of the subject of neural networks had occurred and, with the enormous strides being made in the brain sciences and increasingly powerful computers,

I realized that the time was becoming ripe for even a theorist like myself to become involved in understanding the brain and Mind.

I had already developed a broad-brush explanation of the Mind in the early 1970s, when I had been involved in an earlier version of neural network research. After my return to the same research area more recently, and upon careful reading in the area of consciousness and Mind, I realized that the problem of explaining the 'innerness' or subjective nature of mental experiences was still unsolved. I began to work again on my earlier 'relational mind' theory and used it as a guide to a more neuro-anatomically based development, the 'conscious I'. I will first explain the Relational Theory of Mind (which I developed twenty years ago but published only relatively recently) and develop the 'conscious I' in the following section.

The essence of the idea of the 'relational Mind' is that mental states are produced by comparing incoming sensory activity to similar activity stored in the past and to related emotional evaluations of those past experiences (very likely stored at the same previous time). It is the related past activities which give the depth of meaning or significance to any particular set of inputs. Thus the blue of the sky that I can see if I look up from my writing is so not only because of its intrinsic colour but also because it has the acquired significance for me associated with my earlier experiences of blue skies – lying out on hills looking up at what seemed like the infinite depth above me, sunbathing on a beach in Brittany with the sun beating down, sitting in a deckchair in the gardens of various houses in which I have lived, and so on. It is this set of related past memories, as if activated at a low level, which gives each of us the content of the inner world we carry around. That world is unique to each person, because all the past experiences I have had when I have seen blue skies, or another particular stimulus, will be utterly different from those anyone else has gone through. My world is also very private, because it is hard to discover what experiences I, or anyone else, has had over their lifetimes. One might think of doing so by recording all such events on video, but that would not be enough since there are internal bodily sensations which would be well-nigh impossible to monitor. One might try to discover the nature of the stored experiences by trying to read out the connection weights of the neural networks in the relevant memory regions in the subject's brain. Yet again, that is highly impracticable; even if the weights were known we do not know how all the features of memory are stored nor what form the coding actually takes. My own, and anyone else's, private mental world will remain private for a long time to come.

In spite of the near impossibility of accurately probing a person's mind from the outside in such a manner, there is no reason to give up trying to understand this 'elusive' Mind. In general, the relational explanation allows one to appreciate how Mind could arise from purely physical activity. The comparison between on-going and past experiences would lead to a generally growing conscious Mind as a person develops from infant to child to adult. What to an infant would be a meaningless jumble of edges and corners and blotches of light, to a child would become a 'box' which when touched in a certain way would have moving pictures on one side which would amuse. The young adult might observe the box as an intricacy of modern technology, full of transistors and VLSI circuits which they have learnt about in college. To a much older adult the television set might be an object of derision or of absorption, with complete absence of any detailed understanding of its internal mechanisms as representing one of the advanced gadgets of modern science and technology. The experiences of each age group would be very different in their conscious awareness of the television. Each would be strongly coloured by what they had earlier learned about the set. My claim is that in fact the total mental content of any experience is precisely the set of relations between past and present thereby set up.

This approach allowed me to develop a theory of the significance or the meaning of a new experience. This was given by the amount of relevant past experience that was excited by a present experience. It is difficult to quantify the amount of past experience which was indeed relevant, but one could possibly use the number of similar nerve cells excited by an on-going experience and that induced by past ones. If this so-called degree of overlap were high then the present experience would have high significance, and the conscious state would be a 'full' one. On the other hand, little overlap would correspond to a 'thin' level of consciousness. This latter would correspond to the infant (or the senile person), the former to someone in the prime of life.

The nature of meaning in a given sentence also seemed to be clarified. The statement 'This piece of cake is ill' usually leads to a smile on the lips of the reader or hearer, as I have found when I have read it out during my talks on the Mind. The meaning of the statement is problematic. The words 'this piece of cake' undoubtedly elicit memories of pieces of cake eaten at tea earlier in the day or at home the previous weekend, of the delicious smell of cakes baking when one was a child, of one's own choice of cake, such as rich fruit, and so on. 'Is ill', on the other hand, reminds one of nasty medicine or lying in bed feeling ill, of nurses in starched uniforms, be-stethoscoped doctors, and so on. The overlap between the two sets of memories is nil. They just have nothing in common.

There are two ways to account for the involuntary smile on hearing the phrase. One is in terms of release of tension between warring interpretations, seen for example when the stickleback stands on its head and performs nest-building when it is betwixt the urges of fight or flight. I prefer the second explanation, which uses the predictive wave set up in frontal lobe by hearing the first part 'this piece of cake'. You might expect a range of completions, such as 'is nice', 'is delicious', 'is ready', 'has been nibbled', etc. But no, there follows the quite unexpected description 'is ill'. It is the unexpected, the disagreement with the prediction wave, that leads to the sense of humour. There is also the additional source of humour in 'this' piece of cake, which underlies the speaker's belief in a world of one sick piece of cake, all other pieces of cake being healthy and normal!

In summary, the relational Mind works by comparing ongoing inputs with reactivated memories of previous experiences. This process may allow rapid evaluation of the input. This would appear to confer a decided advantage over animals which cannot use their previous experiences in such a manner. The latter would only have a stereotyped set of responses and would not be able to call on increasing experience to guide them in their contact with the outside world. Even those who did have some storage of previous experiences might be slow in using such stored memories if they did not have a comparison machine like our own. Consciousness would thereby have great survival value for its possessor.

The 'conscious I'

AT THAT POINT in my thinking I realized that the 'relational Mind' was an idea but not yet a theory. It had explanatory promise, but needed implementation in a neuroscientific framework before it could be tested. I began to search for a structural implementation among the midbrain and cortical clumps briefly outlined in Chapter 8. At that time I heard of some of the MEG results I described earlier and became closely involved with some of the technical problems involved. That led me to grasp the importance of a network of cells, called the Nucleus Reticularis Thalami, or NRT for short, which covers the front and sides of the thalamus. All inputs go to the thalamus, through the NRT, and up to the primary cortex. Feedback from cortex to thalamus also passes through the NRT, and all the excitatory fibres passing through it give off collaterals to it. An illustration of the NRT and related thalamic nuclei is shown in Figure 2.

Over at least three decades, neuroscientists have studied the NRT and regarded it as a set of gates controlling access of information into and back out from the cortex. Over a decade ago one of the experts on the NRT remarked:

> Situated like a thin nuclear sheet draped over the lateral and anterior surface of the thalamus, it has been likened to the screen grid interposed between cathode and anode in the triode or pentode vacuum tube.

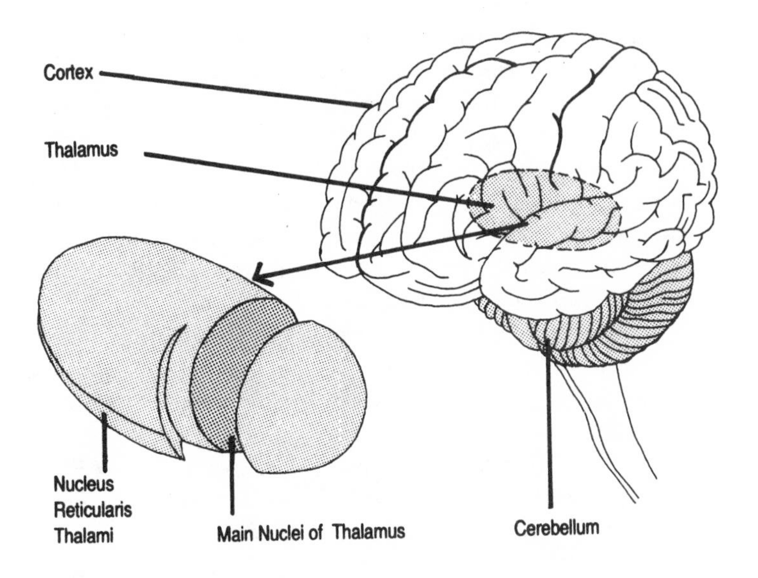

Figure 2: Thalamus and Nucleus Reticularis Thalami (NRT) (from *Principles of Neural Science*, by E. Kandel and J. H. Schwartz, Elsevier, 1985)

More data has since emerged on the detailed structure of the NRT. It has been found to have connections between nearby cells very similar to those of an analogous net, the layer of horizontal cells in the retina which I described in the previous chapter. That similarity gives one interpretation of the 'I' part of the title of this section – the NRT is like an 'eye' looking at the brain. However, the NRT is made up of mutually inhibitory cells, unlike the retinal layer. This would allow the NRT to act most efficiently as the playground for competition between activities in different cortical areas. Imagine two nearby regions of activity on the NRT. Each will try to damp down the activity of the other region by the lateral inhibition between them; whichever is stronger will be expected to win. This has been shown to occur by simulation of a simplified model. Such battles as occur would have to be of a global sort, given the totally connected structure of the NRT, through what are technically termed dendro-dendritic synapses that also occur (as I mentioned earlier) in the retina. These special synapses do not seem to occur, however, in a more primitive animal like the rat, but only in cat, monkey and assumedly in ourselves. I then proposed a theory of the NRT as a global template or gateway to the cortex. The competition is guided by activity from frontal lobe and hippocampus, as occurs in attentional search, or restarted by a sudden input coming in unexpectedly. This latter may well be achieved by the brain stem wiping out current NRT activity. Analysis of the appropriate mathematical equations, similar to those I used for modelling the retina, and also computer simulation, has indicated that control of activity over the cortex was indeed exercised by the NRT over a long range. Moreover the NRT sheet could support a backward-moving wave of activity, provided that its connectivity was suitable. This and numerous other specific features made me realize that the NRT could function, along with its nearby cortical areas and thalamic nuclei, as a global comparison network for activity over the cortex. The 'conscious I' approach to the NRT may be an important component of consciousness.

One of the crucial features of the ongoing comparisons being carried out was the ability of the system to compare incoming activity with stored past experiences. It turns out that the front part of the NRT sheet seems to be connected so as to have such experience from previous memories imprinted on it. The limbic circuit, mentioned in the previous chapter, which stores past experiences and their emotional significance, sends its output directly through this front part of the NRT to the thalamus. The latter does not appear to return the compliment, so that it (and so consciousness) exercises little immediate control over these memories, or over emotional and drive features of brain activity. Only after some cortical processing will the conscious activity filter through into memory or emotional centres – and so have later effect. Consciousness is thus the relational structure set up by the multiple gates of the NRT, all correlated by the lateral inhibition intrinsic to the action of the NRT sheet, together with the correlated activity in cortex and thalamus. Cortical input filters down to the memory and emotional centres, which give colour and content to the expectancy wave fed back from activity imposed on the front part of the NRT sheet, and which helps determine allowed cortical activity. This is the origin of the expectancy wave. It meets input activity, and the relation between the two determines future conscious states.

There is still an important feature of the model which needs amplification. By the 'Relational Mind' theory, the relevant memories for a given input give the overall conscious content, the allowed memories being determined by the 'conscious I' activity (very likely controlled by the NRT). But the detailed content of these memories has to be fleshed out to give the inner content of conscious experience. That would probably occur in a manner similar, but at a lower level, to imagery. In imagery past memories can be used to re-create, in the 'mind's eye', some pictorial features of previous visual experiences (with similar re-creation of memories in other modalities). Various measurements of brain activity (especially from increased blood flow in active brain regions) have shown that during imagery regions of brain near the primary input regions are being used. This is to be expected from what was described in the previous two chapters, since these are precisely the regions used in the analysis of inputs. It is reasonable to expect that these same regions are used in the re-creation of the specific features that occurred in the earlier experiences, so as to make the images look similar to input-driven experiences.

In consciousness something like the images of appropriate memories will very likely be re-created in brain areas close to input regions. The most effective way for them to combine suitably with the input activity (and thereby avoid hallucinations) would be by means of persisting input acting as a 'seed' for the imagery feedback. This 'seed' could well be equated with short-term memory, which is the ability to hold an object in memory for up to several seconds. There are many feed-back wires from long-term memory storage regions in the temporal lobe back to the earlier 'image-creation' regions. These wires could thus feed back allowed memories preferentially to short-term memory of input, so fleshing out the consciousness of such input. In this context it is important to note, as pointed out recently by Francis Crick (one of the discoverers of the genetic code) and a colleague, that there is no case known of loss of short-term memory when consciousness persists. Thus we can arrive at the interpretation

consciousness = imagery of allowed memories, controlled by the 'conscious I' complex.

This model allows one to understand the uniqueness of consciousness and the emergence of two separate consciousnesses in split-brain patients, which occurs when the band of fibres joining the two halves of the cortex is cut so as to reduce otherwise intractable epilepsy. A subject, 'P.S.', who had this operation, emerged with two separate conscious hemispheres having different likes and desires.

Higher levels

THE 'CONSCIOUS I' model of consciousness has implications for drives, emotions and higher cognitive processing such as planning and thinking. We considered drives and emotions briefly in the last section, where the lack of conscious control of emotions was explained by the details of the appropriate wiring diagram. Planning and rational thinking were noted in the previous chapter as being crucially involved with the frontal lobe. That has inputs and outputs to the so-called medio-dorsal nucleus of the thalamus (MD for short), which has contact with the front of the NRT. This front region of the NRT was proposed as the crucial one where competition for consciousness is being fought out by emotions and cognitive experience. But then the relevant parts of the frontal lobe will also be involved in such battles for conscious control; planning and reasoning will have direct access to consciousness. The manner in which unconscious processing occurs, in particular during creative activity, and the emotional colouring of conscious experience, would seem to be possible to explore by investigating further details of the connections between thalamus, the frontal lobe and the limbic circuit. This would also allow further development in understanding drives and goals.

An important part of the 'conscious I' model is the view it gives on sleep and dreams. In the former, it is well known that the thalamus and the NRT function in a manner which both cuts off the cortex from all inputs and prevents it from functioning normally (as evinced by the corresponding slow cortical EEG rhythm). In dreaming there is a similar cut-off. Yet dreams, in spite of being somewhat conscious, are out of conscious control. As noted by Eccles:

> A characteristic feature of most dreams is that the subject of the dreams feels a most disturbing impotence. He is immersed in the dream experience, but feels a frustrating inability to take any desired action. Of course he is acting in the dream but with the experience that in doing so he is a puppet.

This can be understood in terms of a possible reduction of the global control by the NRT of an active cortex (which has fast EEG activity similar to that for an awake person). Dreaming can be termed 'cortical consciousness', while the awake conscious state can be called 'thalamo-NRT-cortical consciousness'. The reduced involvement of the NRT in dreaming may help explain the feeling of impotence noted by Eccles. There is clearly much to be understood about dreams, but at least we might here, in variations of the 'conscious I' model, have a guide.

What about the self? The simplest explanation is that the concept of 'self' is possible to humans because we have enough cortical capacity to build a model of ourselves as actors. This only begins to develop in the infant from the age of eighteen months or so. In terms of the ability to model experience by memory storage in neural networks, the problem of building a neural model of the self would only seem to be on a par with that of modelling of other experiences. That we are able to have a self-model gives us an immense edge over other animals, because it allows us to experience the 'intentionality' of others as if they were us. Communication is different when the intentions of the sender of a message resonate chords of similarity in the brain of the recipient. We may modify the identity at the end of the previous section to

self-consciousness = imagery of self-memory, controlled by the 'conscious I' complex.

To claim to have explained the self in one paragraph and the mind in one chapter is clearly absurd. This account – the 'conscious I' – is a tentative step towards an underlying mechanical model. Both it and the earlier 'relational mind' are presented to show how the problem of Mind may be tackled. The ideas expressed here are in harmony with other ideas of the Mind such as the 'global workspace'. In this model it is suggested that Mind is essentially a theatre in which to make public the activity of various sensory processors. Moreover many neuroscientists agree that the NRT plays an important function in brain processing, and consciousness is the result of a complex of brain activity. The 'conscious I' develops that into a model in which the NRT is 'looking' at the cortex and thalamus and allowing global competition to give a unique winner. Whether or not the details are correct, I claim that consciousness will not look the same again, and people will appear in a very different light as magnificently subtle machines.

PART FOUR

UNDERSTANDING THE UNIVERSE

Chapter Eleven THE ANSWERS

The impossible questions

THE MAIN CONTENT of the book so far has been to attempt to give answers, based on science, to the four impossible questions:

How can we explain consciousness?
How was the Universe created?
Why is any TOE that particular one?
Why is there something rather than nothing?

Before I consider the relevance of these answers to the human condition, let me summarize in this chapter the answers I have given earlier.

Explaining consciousness

LET ME START with where we have just finished. The basic thesis was that the phenomenon of consciousness arises purely from brain activity. The 'mental content' of conscious experience is constructed, in terms of the Relational Theory of Mind, as the set of relations between on-going brain activity and that aroused by this activity from stored memories of related earlier activity. Such relations are not material objects themselves (in the same way that comparisons between numbers, such as the fact that two is less than four, are not numbers) but give the inner detailed mental content of conscious experience.

The relevant relations to give this mental content are set up by activation of memories suitably close to the input activity and so require a comparator or 'gate' to achieve them. It was suggested that this comparator consists of a sheet of inhibitory neurons placed at the entrance of all inputs to the cerebral cortices, and is called, as we noted above, the Nucleus Reticularis Thalami. Activity on the NRT undergoes global competition so that certain activity on part of it ultimately dominates. Recent experimental results were used to suggest that the total activity on the sheet can be controlled by that on only a small part of it. Attention controlled by higher cognitive processes is posited as occurring when these processes send their dominating activity to the front part of the NRT. This sweeps backwards, so that activity on the rest of the NRT and on related areas of the cortex are consequently controlled by it. Novel inputs draw attention to themselves by destroying the on-going NRT activity and letting it start its competition afresh. Consciousness itself is suggested as arising from the comparisons set up between the activity impressed on the front part of the NRT (which was excited by earlier inputs or from frontal lobe) and that coming in from input through the primary sensory regions of the cortex. Emotions and drives are seen in this framework as playing a similar role to cognition, fighting to be the winner on the front part of the NRT sheet. The lack of direct conscious control over brain activity in areas which are not in close enough contact with the NRT explains the difficulty we all can have with our emotions – 'uncontrollable urges', 'blind rage' and so on are quite accurate descriptions of such lack of control. Finally, self-awareness, a faculty possessed by humans, chimpanzees and orang-utans, but very probably by only a few other species of animals, is posited as arising from use of the consciousness machinery in combination with an internally stored representation of one's self. This latter is expected to be only possible provided there is enough spare capacity to allow storage of the myriad facts of one's own personal life – visual appearance, aches and pains, likes and dislikes, abilities and weaknesses, and so on. The details of how this storage is achieved and accessed is still unknown, although there are numerous ways in which it is thought this could be achieved using recent developments in neurocomputing and artificial neural networks. In all, the broad outlines of the underlying mechanism of human consciousness seem to be emerging and the problem of obtaining Mind from Matter solvable even at a quite detailed level.

Creation of the Universe

THE SECOND impossible question was how the Universe was created. Two possible scenarios were presented. The first had the Universe as 'popping out' of the vacuum as a fluctuation of the quantum state, like the foam (analogous to the fluctuation) arising on the surface of a wave (the vacuum) on the sea. This was seen to have an equivalent difficulty to that of setting up the Universe as a clockwork machine, where a decision was necessary as to the form of the initial dispositions of all the particles (their positions, velocities, etc.). Who was supposed to be present to make that vital decision? The analogous problem for the quantum creation of the Universe was the choice of the initial state of the Universe or, more technically, the initial wave function. Various recipes for making such a choice were described, and each was seen as shifting the responsibility but not removing it all together. It was noted earlier that the present technology for analysing the implications of any such recipe for the choice of initial wave function is woefully inadequate, and that claims of solving the problem by use of imaginary time or the unobservability of high gravitational fields were both founded on an explicitly incorrect theory of quantum gravity. More importantly, however, it was realized as crystal clear that no unique recipe is ever going to be available for singling out one initial state from another, so the possibility of explaining the creation of the Universe in this manner is seen to be hopeless. The only feasible scenario for the creation of the Universe that avoids a vital arbitrary decision (choosing one recipe rather than another) was then explored. It requires the existence of an infinite regress of theories, each giving a better description of the Universe as its temperature rose and its time from creation decreased. At each stage of the progression back to the beginning, in which the description at one level had to be replaced by that at the next higher level, the initial state of the Universe at the lower level could be expected to be deduced from that at the next one up. If one knew the wave function of the Universe when quantum mechanics was appropriate, then the classical distribution of particles defining the initial classical state of the Universe could thereby be calculated.

The use of an infinite regress of this sort always allows the initial stage of the Universe at any one level to be determined from its state at a higher level. Yet there is no point in searching for a unique recipe to specify that state at any level, since the theory at that level will itself ultimately be found to be an incorrect description of the Universe at a suitably higher temperature. Experiment must be used to guide the search for correct descriptions. The choice of initial state at a certain level will determine the future nature of the Universe at all later times. Guided in addition by certain helpful criteria (though ultimately unjustified), such as elegance, symmetry and so on, the initial state at that level can be determined, and so act as a constraint on the initial state at the next higher level. In this manner, the beginning of the Universe (and the ultimate initial state of the Universe) recedes into the infinite past; it is never accessible, although we should always be able to progress towards it.

The impossible TOE

I HAVE JUST explained how much more natural it is to try to explain the creation of the Universe in terms of an infinite regress, never quite getting to it but always describing it more accurately. Science, as a probe of the reality around us, makes ever more accurate statements about it, but the theories used as the basis for these statements are always prone to modification. On the evidence we have as a result of over four hundred years of research and experimentation on the physical world there is no reason why these modifications will not always occur; there will be an infinite sequence of theories, sometimes branching, sometimes fusing to give a single theory (but this will always be changing).

If the above is correct, then there can be no theory of everything. The basic problem which I described for any TOE was: why does it take the form it does? If it really is a TOE there would be no way to explain its basic axioms. Moreover, if it were correct experimentally then there would appear to be no possibility of explaining the properties of the 'fundamental particles' whose existence would be required by the TOE and which presumably would be observable by suitable experimentation.

Can all phenomena be explained? This is the question at the heart of the nature of existence. If the anwer is 'yes' then we are required to have an infinite regress. Reductionism will be right but it will never terminate. In the words of a poet: 'Big fleas have little fleas, upon their backs to bite 'em, Little fleas have lesser fleas, and so *ad infinitum.*' If the answer is 'no' then science will fail. However, we can only find the scientific answer to whether or not it is possible to give an explanation of all phenomena by experimental investigation. This would require positing the nature of some underlying, as yet unobserved, substructure for the regularity observed in nature. This substructure would allow experiments to be developed and performed to test the assumed further theory corresponding to the substructure.

That is not to denigrate serendipity and random search in science, which has led to highly important advances such as the discovery of the microwave background radiation bathing the Universe, or the new generation of high-temperature superconducting materials. Yet fundamental scientific progress will still only be made by probing further the very large or the very small, and that will continually require ever more expensive equipment. From a practical point of view, it will be increasingly difficult to find the financial support for experiments mounted just as 'shots in the dark'. From a creative point of view, it is highly unlikely that scientists would stop speculating about the nature of any so-called TOE claimed by some of their colleagues, and one might even expect this to be regarded as an excellent challenge to their creativity and ingenuity. The scientific response to the claim of any theory to be a TOE will therefore have to be of disbelief and the assumption that fundamental science has not reached its resting place. 'We never close' must be the motto, for if we did how do we know that the next experiment or the next idea would not bring down the TOE edifice claimed to be the end of fundamental science?

I conclude that there is no way of justifying the statement that a particular scientific theory is a TOE. The theory will always be vulnerable to recalcitrant phenomena, like a coconut waiting to be knocked off its stand in a fairground. Moreover, present evidence indicates that there will, after all, be an infinite regress; new theories will continually be discovered and science will never come to an end.

Something rather than nothing?

THE FINAL impossible question was noted as unanswerable scientifically in the framework of the infinite regress approach. Let us suppose that one has deduced, for some reason or another but on the basis of a given theory, that a particular set of objects with certain properties had to exist. The theory is ultimately doomed to be destroyed by a better theory. At the same time our necessarily existent objects would have lost their warranty of existence. However, we have no guarantee that more precisely defined objects exist at the more exact level. Thus proof of existence of anything at any level is unsatisfactory since it will only be useful transitorily.

The ultimate nature of reality appears to disappear before our gaze when we accept the possibility of an infinite regress of scientific theories. We will never know in its entirety the exact nature of any object in the Universe. In no way should we interpret this as implying that reality is any less precise than we had thought before, without the possibility of an infinite regress. In our progression of scientific theories each one is ever more accurate in describing phenomena around us. It is just that an infinite regress makes it infinitely more difficult to discover that reality. Nor is this a message of pessimism about the loss of perfect clarity. On the contrary, it gives us optimism that we can expect to look at the world with ever greater clarity, commensurate with the effort we put into it. We are like a poorly sighted person who can see increasingly better provided they get better spectacles. It will be more and more costly for them to see more clearly. Only if money were no object would they be able to see ever better.

Beyond science?

WE HAVE NOW explored some of the features by which science is in principle bounded when used in attempting to understand the Universe. They are not limits to scientific knowledge (which appears unlimited) but only to the ability to answer certain of the impossible questions. In particular the likely infinite regress of scientific theories will always hide from our eyes the ultimate nature of reality and the reason for existence. What, then, have we gained from science? The approach to the impossible questions taken by religious faith was criticized at the beginning of the book as having no firm grounding in reality, nor giving much in the way of detail. Yet science does not seem to have come up with any definite answer either. It could be argued that there has not been any clear improvement in the situation. Indeed it might even be claimed that we have been forced to take steps back from the supposed certainty supplied by religion if we follow the scientific route. All certainty has been lost, dissolved before our eyes.

That response can be answered on two fronts. The first involves the method of approach which is claimed by believers to lead to religious revelation and to be at the root of the nature of

God. It is in the mental world that God is supposed to exist, and through prayer and religious experience of a mental form that knowledge of God and communication with Him is made possible. I have described very strong evidence for the claim that the mental world is itself purely the result of activities of myriads of interconnected nerve cells, albeit of a very subtle form, and that there is no extension of that world for each of us beyond the confines of our brain. In accordance with this, there seems to be no truth in claims being made that religious experience is a way of finding certain knowledge, and in particular of discovering a mental Universe distinct from the physical one. All such claims are delusions, perhaps based on the emotional needs of the perceiver. The only real experiences of such percipients are those of altered states of consciousness, ultimately explicable in purely physical terms. Such states will have no greater importance for gaining real knowledge of the Universe than do drug-induced ecstasies. The latter can lead to remarkable alterations in visual perceptions, such as seeing beautifully symmetric shapes. It turns out that these phenomena are all explicable in terms of wave-like activity patterns which can be created in certain neural networks under suitable conditions. The effect of drug-distorted information-processing at higher levels produces emotional distortion, with altered significance attached to the surroundings, as compared to when no drugs are taken. These experiences can be expected to have a similar explanation as in the case of drug-induced distortion of the purely visual experiences. There is no reason why religious revelation should not have a similar explanation, although now the drug cause of the altered states must be replaced by emotional disturbances or drive states bringing this about. It was noted that emotions are not under direct conscious control, and will be a powerful influence on much behaviour. Their direct production of religious revelatory states would be unsurprising.

There is also no foundation to the charge that science has not led to the discovery of any better certainty than has religion. There are no limits to science; any phenomenon can be explored scientifically *ad infinitum*. No answer is ever accepted as final and there is no other arbiter of the truth except for experience itself. But that experience must be seen as part of the set of all experiences, and explained in a testable framework of scientific understanding which embraces as much of nature as is possible. Any other way is far less systematic and justifiable. This is especially so for religion. It claims certainty about the nature of the mental world, but never correlates that with the physical world, nor gives any detailed understanding of that mental world. Moreover, one could question strongly the nature of religious certainty. God is the ultimate arbiter, but existence of the enormous divergence of religious views and of differences of opinion even in the interpretation of any one religious sacred book (such as the Bible or the Koran) indicates that God's messages are not being received very clearly. Because there are so many interpretations one must question the reality of the message. In contradistinction to this, the scientific message is unanimous. When recalcitrant phenomena are discovered which do not fit in with the current theory, there is usually a rearguard action by the establishment to hold on to the status quo. This is done, it would seem, mainly because the old guard had already invested much time and effort in establishing the now orthodox view, very likely against stiff opposition from the previous establishment which they had displaced. Such opposition to too rapid change is necessary in science – new results or ideas might be wrong and have to be checked very carefully before being accepted as part of the established corpus of science. But then unanimity descends once again.

For these two sets of reasons – the ultimately physical nature of religious revelatory experience, and the unanimity of science compared to the vast spread of religions – science can be claimed to provide us with an increasingly certain picture of the Universe and of ourselves within that Universe. However, the scientific path has apparently not helped us towards an answer to the question 'Why is it all here?' It is the answer to that question which we hope to use to guide us in our lives. If we could see purpose in the Universe at large then we might see it in our own lives. If the Universe could be shown to exist in order to have the greatest 'pleasure' or 'harmony', or some such emotionally charged purposes, then science would have been eminently helpful and we would become hedonists or 'harmonists'.

Science would not be expected ever to lead to observation of our emotional qualities echoed in the Universe at large. For we have only ever observed these characteristics being exemplified in living systems with the very special structures of sensitive neural networks. These can only exist under specific conditions in the Universe. Conscious living neural networks are only recent comers in the Universe (existing for the last sixty million years – life having been on earth for over 3500 million years), and then only in very special places and environments which support their delicate working.

Not only do we not expect to see any emotional traits in the Universe at large but we do not expect to see purpose. There is an exception to this in that all modern scientific theories can be written in a form in which there is some quantity (such as an energy) which tries to attain an optimum value. This means that there is the limited purpose in the Universe of optimizing that particular quantity, but to use that to guide our own lives would seem impossible. We are microscopic sub-units in the vast total of the Universe. We could try to hold our own little corner by optimizing whatever parts of the total Universal quantity is possible; our solar system is relatively isolated from nearby stars in the Milky Way, for example. But the conservation of energy which this optimization would correspond to is in any case one of our assumed basic laws of physics. As I have said, there is no humanity in the conservation of energy, no emotions, ethics or morals. Our mental life has arisen because we were in the right place (Earth) at the right time (as mammals evolved) to respond most effectively to survival pressures to adapt by becoming conscious. If our mental world evolved by earlier survival pressures then we must note those pressures now and ask how best to continue surviving.

Survival of the intelligent

THE PRESSURES to survive in the last few million years, when the human brain expanded in size most rapidly, were those of competing in an uncertain and ever hostile environment with other predators. Since those other predators included other humans, it was a question of pitting brain against brain. The humans with more complex brains survived better because undoubtedly they were more flexible, had better memories and could develop better strategies for catching prey than their dimmer-witted fellows. But we in the developed world no longer live in caves and rarely hunt wild animals other than for sport. Technology and medical science have moved on so far that evolution has been overtaken by genetic engineering, and alien environments have been replaced – air-conditioned living in desert conditions and centrally heated luxury in the cold depths of winter. Electric light has allowed us to turn night into day. Altogether life has become very different for *Homo sapiens* from that of his earlier hominid relatives.

We now share our own destiny with each other in our increasingly interconnected society, and must look to see how scientific understanding can act like a refiner's fire on our modes of living. That we explore in the final chapter.

Chapter Twelve

THE IMPLICATIONS

What does it all mean?

I BEGAN THIS book by raising the question heading this section together with related ones – 'Who am I?', 'Is there life after death?' – and similar questions of a personal nature. I then indicated that the most effective way to answer these questions was through science, leading thereby to the deeper 'impossible' questions: 'How can we explain consciousness?', 'How was the Universe created?', 'Why is any TOE that particular one?' and 'Why is there something rather than nothing?'

The first of these 'impossible' questions was given a tentative answer in terms of a Relational Theory of Mind. The relations were comparisons set up between stored memories and present experience by means of a special network of cells situated between the sensory organs and cortex. This approach results in the Mind being seen as purely a creation of material activity. It is a materialist theory of Mind, yet in no way denigrates the importance of consciousness and self-awareness in granting great powers to the possessor.

The second and third 'impossible' questions were responded to in what was concluded as the most effective and scientifically realistic model, that of the infinite regress of ever more precise scientific theories. There would therefore be no TOE, nor any final answer as to how the Universe was created. Nor could any answer be given to the final 'impossible' question, since any explanation would have to be couched in terms of a scientific theory, at a given level, which would become defunct sooner or later.

It would appear that no meaning at all can be given to any life form, other than that it appeared because the conditions were favourable for it to do so. Life is no different from non-living Matter as far as the Universe and its laws are concerned. Human life possesses the almost unique power of self-awareness, apparently arising as a result of evolutionary pressures. Nothing exists in the Universe that is other than physical, albeit with the ability to produce decidedly non-trivial behaviour such as that of Mind. Having thereby discarded Mind as giving any direct access to God, or to any other entity supposedly able to give meaning to life, we now reach the position that the true nature of reality – by which I mean physical reality – can never be seen in its entirety.

In any case, human desires, goals and emotions are seen in this framework to be irrelevant to the infinite chain of scientific theories which will forever beckon us enticingly forward in our scientific quest to understand the Universe better. These human features can only be investigated as part of the developing scientific study of consciousness and emotions.

Life without meaning

IT MIGHT BE claimed that one or more of the answers to the impossible questions which I have arrived at are wrong. In particular, the Relational Theory of Mind has not been scientifically proved and a great deal of work will have to be done to show it is valid. Yet I am sure that it, or a theory very like it, will be validated over the next few decades. The spotlight of scientific curiosity has now turned to consciousness. The degree of interest in it – with several conferences this year alone and a new scientific journal devoted to it – shows that the time is becoming ripe for great strides to be made. I further claim that the Relational and 'Conscious I' theories which I have presented have a good founding in neurobiology and lead to new insights into both consciousness and emotions. They may even lead us to comprehend the nature of consciousness in animals – cats, dogs, horses, etc. – in a way previously impossible, as well as helping people with mental disorders. In any case they give a scientific theory, which can be tested and improved (or rejected) by experiment in the near future.

Moreover, the infinite-regress scenario of science is only a formalization of what is already seen as a trend in science, being the other (and more likely) alternative to the TOE possibility. It allows causality – no effect before cause – to continue its triumphant progress as a well-attested principle into the future. There are therefore overwhelming arguments for what it is appropriate to label the 'materialistic, infinitely regressive' view of the Universe that I have presented here.

We now have to face up to the challenge in front of us: how do we live our lives if no deep meaning can ever be found in the Universe? That is the ultimate challenge each of us faces when we have inwardly digested and thought through the facts and theories relating to the Universe which I have developed in this book. It is one to which I myself must respond.

The supposedly bleak message carried by this view of humankind and the Universe is anathema to some. They claim that there must be a purpose in life, so turn away from science and may embrace a religious faith. Yet the subsequent decades and centuries will see the ever further encroachment of science on such faith. Technological devices already make our lives easier. They will increase in numbers and power. As just one example, the 'phonetic' typewriter, which produces type directly from spoken material, is already on the market in special versions and will ultimately be as mass-produced as the personal computer or Walkman. But it will be the development of devices with a modicum of consciousness (initially so-called 'attentional computers', but later those with consciousness and emotions), which will make the religious position ultimately untenable. There are already research groups producing machines with limited capabilities in these directions. The retreat from reason to faith will become ever more difficult as such understanding increases. In the end, consciousness and Mind will be universally accepted as having been explained materialistically in the same way as the atom or the stars in the heavens have been. It may take a thousand years to achieve such a level of knowledge, though with the present speed of scientific advances in brain research I would give it only a tenth or so of that time. But that will be a minuscule fraction of the time mankind will still have on earth before the sun's certain death throes as a supernova some billions of years hence. How will humanity face up to the apparent meaninglessness of life during that time?

Problems in society

CONSIDER THE following events:

> On 15 July 1992 a young mother walking her two-year-old child across a beauty spot on Wimbledon Common in south-west London was sexually assaulted and then brutally murdered by having her throat cut. The child, caked in mud and blood and beaten around the head by his mother's killer, was found clinging to his mother's body.

> On 16 July 1992 hundreds of youths stoned police, burnt and looted shops on a council housing estate in Bristol after two young men who had stolen a powerful police motor cycle had subsequently been killed when they ran into a stationary police car. As *The Times* reported: 'Among the crowd were women with children in pushchairs and toddlers revelling in the antagonism. At one point a sweet shop was looted. Some of those involved were said to be as young as four or five. Parents on the estate made little effort to keep youngsters indoors, allowing them to watch police in riot gear patrolling the estate.'

> In April 1991, in Los Angeles, thousands of people rioted and caused a number of deaths by shooting. Shops and homes were burned over a period of a week, causing the National Guard to be called out and a curfew to be declared. This was

the worst riot in the USA since the Civil War in the previous century.

These are but three examples of man's inhumanity to man in the so-called 'civilized' Western world. The first two cases could be multiplied many times. The death and suffering in full-scale wars, particularly the carnage of the two World Wars, but also in numerous smaller ones causing further anguish, make this century appear one of the bloodiest and most brutal. Yet massacre and suffering have left their toll through the ages. Violent death was common, with flaying alive, crucifixion and burning at the stake carried out. Human life may have been adjudged less important in the past, yet terrible human pain and suffering were experienced none the less.

It would seem that whatever guidance has been available in the past has been rather ineffective in certain instances. Religion, supposedly the provider of a code of morality, did not prevent violence being perpetrated under the religious mantle. The Inquisition and the 'holy wars' are reminders of past religious-inspired iniquity; the 'fatwah' of Islam and the 'just war' blessed by Christian bishops are with us today.

On the other hand, religion has guided countless many to live in a manner causing little distress to those around them. These examples indicate that there is still the strong claim to the good life if lived, say, by the Ten Commandments. But why live such a life if ultimately it has to be accepted that there is no meaning to it? Why should one not live so as to get as much enjoyment from it as possible, and take no care for the consequences of one's selfish actions on others? If one is unable to do that in a peaceful manner might one not even be justified in trying to take forcibly from others so as to live up to one's expectations? Indeed the welfare-state mentality seems to support the position that the State has a duty to supply the material for a decent standard of living and is in default if it does not do so. Hence looting and pillaging of others' goods is held by some to be partly justified, as in the Bristol and Los Angeles riots. The poor murdered young mother on Wimbledon Common is a victim of emotions run riot and the inability of society to deal with murderous emotional outbursts.

Nothing that I have written so far can help support the antisocial activities described above. Nor does it lead to the out-and-out hedonistic approach to life by just searching solely for one's own self-gratification and the devil take the hindmost. Unless one lives in total isolation from society, in a completely self-sufficient manner, one has to interact with others. In that interchange much can be gained (or lost), good or bad, by the individual in his manner of approach. If a person were to go around being rude and selfish to others (or even sexually assault and murder if the need arose), they would not themselves get very friendly responses nor be given help in the many ways needed in the modern world. Their life would then become progressively more difficult. It is possible to see the beginnings of a strategy for response which avoids the pitfall of heaping up problems for oneself by always being selfish. It is one which has been increasingly investigated with scientific precision over the last few years, and it has led to a surprising result.

Nice guys come in first

It used to be said that 'nice guys come in last'. Yet it has been found quite recently that in a society made up of people who could respond either in a nice or a nasty way to each other it is the nice guys who will tend to come in first. This unexpected feature was noted in a gambling game called the Prisoner's Dilemma that is played between two players who each have two cards or choices, one called 'co-operate' and the other 'defect'. Each player plays one of his cards face down on the table. The cards are then turned over. The banker (someone other than the two players) pays a reward of, say, £300 to each player if they both play the 'co-operate' cards, but fines them each £10 if they both 'defect'. On the other hand, if one co-operates and the other defects the one who has defected is given £500 as a reward for defecting, while the other is punished by a £100 fine for not having defected. If the players had actually been prisoners, the fines and rewards would have been longer or shorter prison sentences for having co-operated with their partner and not 'split', or having turned Queen's evidence and told on their partner (equivalent to handing in the defect card).

The correct strategy to play is clear. Suppose we are playing each other. If you play your defect card, then I must do the same so as to lessen my losses (which would then be £10 instead of the £100 fine I would have suffered if I had played my co-operate card). If you play your co-operate card I could gain £500 if I defect, instead of only £300 if I co-operate. In both cases I should play my defect card. Two rational players, as I assume we both are, would then each lose £10. Yet the dilemma is that if we had both co-operated we would each have gained £300. Since there is no way of ensuring trust between each other in the game then we each lose a small sum.

Now consider playing a number of games of the Prisoner's Dilemma, one after the other. If I defect each time, and you are either very silly or suicidally altruistic and co-operate in every game, then I would win, say, £5000 in ten games; and you would lose £1000. But it would take a saint or a madman to continue trying to co-operate with me if I behaved so badly. Alternatively, we could play each other in a co-operative manner throughout and each gain £3000 after ten games (with the poor banker losing that amount in the process).

It seems that there are numerous species which do play such a strategy, by which they gain through mutual co-operation. Thus about fifty species of small fish and shrimps make their livelihood by picking parasites off the scales of larger fish. These latter benefit by being cleaned and the 'cleaner fish' by obtaining food (the parasites themselves) in this way. While a large fish could have gobbled up a smaller one after it had been cleaned that does not seem to happen. This is quite remarkable, especially when large fish sometimes open their mouths and allow their teeth to be cleaned by the smaller fish. It would seem reasonable to suppose that this altruism gives evolutionary advantage to species.

The idea that there may be better strategies than straight nastiness (always defecting), to allow some sort of adaptive response to the other's previous play, has been tested by taking a range of alternative strategies, including 'always defect'. Pairs of players, using different strategies from this range, then play against each other. It was discovered that the best strategy for winning, when the score is averaged over a set of games, was a forgiving one called Tit for Tat. This consisted of a player only retaliating after being provoked by the other player defecting. If the other player stops such dirty play and co-operates then the tit-for-tat player does so also. It turns out that if many pairs of individuals can play against each other and the winning strategy is paid a reward not of money but of offspring identical in strategy to the parent, then the Tit-for-Tat players eventually become more numerous than those using other strategies after about a thousand generations (provided the Tit-for-Tat strategists start with enough players on their side).

This, and similar results on the Prisoner's Dilemma, has been arousing a lot of interest in trying to understand how evolution could have led to altruistic relationships between, for example, large fish and cleaner fish. It has also been used in the past in strategic planning in war games. But it should, and I have no doubt will, be important in helping guide us as to how to live our lives most effectively in the modern world.

A strategy for living

IN SOCIETY WE have, and indeed must have, constant interactions with each other. In a modern world in which science and technology can be increasingly regarded as the banker, providing us with ever-increasing rewards, the winning strategy for the repeated Prisoner's Dilemma will be Tit for Tat. That means do not strike out first, but co-operate wherever possible with others. In some ways that gives a moral code similar to the Ten Commandments. The 'tit' retributive step, in response to a nasty 'tat' given by an unpleasant player, has to be handed over to society and is embodied in its laws and penalties. To take the law into one's own hands would destroy the fabric of modern society.

The justification of the Ten Commandments and of laws and penalties against those who transgress in society, on the grounds of ideas and theories discovered by analysing the development of colonies of artificial beings through evolutionary ideas, seems at first somewhat difficult to defend. For modern society would now claim to have moved beyond the control of evolutionary pressures. Science and technology have allowed us to take our destiny into our own hands, and even to create new species by genetic engineering. Yet the ability to change ourselves so as to reduce our aggressive tendencies appears somewhat remote in the reasonably short term. It is more critical that we learn how best to live with what we have genetically at hand, namely mankind with its present genetic structure. The discovery that the tit-for-tat strategy is so stable indicates that we should attempt, each and every one of us, to use this strategy if possible. We must therefore use not genetic but educative methods to enable humanity to live more effectively together in the reasonably short space of time allotted to each of us.

But is this not where we came in earlier, with the accounts of violence – the brutal killing of the young woman, the looting and killing in streets of towns and cities? We still have to discover how to persuade people to help each other in a world full of deprivation and alienation. The home and school seem to be failing to achieve this. Too many teenagers reaching adulthood are almost illiterate and just about unemployable. They have been brought up on a daily diet of television and videos showing scenes of explicit sex and violence. They have had little moral guidance from their parents and not enough education from their teachers. No wonder there is so much violence and unpleasantness in society today. Nor is the scientific advance entirely blameless, since new techniques such as automation have led to increasing unemployment.

It is clear that much improvement is needed. Stronger codes of ethics as to what is allowed to be played on a television set should be imposed. But they would be very difficult to enforce. However, education could still be a powerful force for good. One indictment of the modern form of education is the concentration on the intellectual, to the detriment of the emotional life of the pupil. Emotional education should also be a part of the school curriculum. It cannot be left to chance, since it could well end up being left to 'sex, lies and videotapes'. The neglect of emotions goes back to the classical Greeks, who regarded emotion as unclean and involving man's 'baser' nature. Yet our lives are only given colour by our emotions; they can no longer be neglected in such a lop-sided manner. As the emotions are better understood over the next centuries, as part of the advance in understanding consciousness, humanity will undoubtedly be able to develop techniques which extend, say, the emotional memory exercises developed by Stanislavski in method acting.

Konstantin Stanislavski was a theatrical producer and director and a founder of the Moscow Arts Theatre in 1898. He developed a new approach to acting based on realism and the actor's complete identification with his character. This idea, later translated to various 'method' acting studios in New York and inspiring actors like Marlon Brando, Paul Newman and James Dean, used various techniques to allow an actor to enhance and control his past emotional experiences so as to bring them to his role. One of these, called 'emotional memory', required the actor to recreate from memory particularly moving experiences from his earlier life. These were then re-enacted in front of the teacher, the protagonists being imagined as peopling the various parts of the studio set. Questions were asked afterwards to probe further as to the details, and the actor would also work separately on making his remembrances more precise. In this manner the emotional sources of an actor's actions were revealed and made more controllable. They were then put into action in improvisation pieces and study of small scenes from a range of plays. Practice at visualization of imaginary objects and their manipulation were developed in other technique classes. If classes like these in emotional memory, technique and improvisation, and incorporating newer knowlege being discovered in psychology and brain research, were developed and used in schools, especially at the secondary school level, it is possible that the emotional frustrations experienced in adolescence could be ameliorated, and later adult excesses (such as the Wimbledon killing) be reduced. Of course, this would have to be done with great care so as not to lead to young people being so overwhelmed by their troubles as to be driven at one extreme to suicide or at the other to inertia. One could also consider a class of children playing the Prisoner's Dilemma and discovering who wins. By classes in simple games theory and by improvising in small scenes around the games theory they had learnt, they could see how to control their lives by more effective strategies than 'always defect'. Then slowly, over the decades and centuries, humanity might become more humane. In the process, enjoyment of life would come through more effective social contact and through creative activity using both developed emotional and intellectual talents. Life would be enriched by the co-operative nature in which it is engaged.

A future full of emotionally and intellectually well-educated people, with science and technology under their power to produce all the necessities of life with little labour, has been a utopia which visionaries have constantly conjured up but which has not yet come to pass. Science has been seen as unleashing powers which are hard to control, but that may be because we have been restricted in our thinking by doctrinaire answers. But as doctrine retreats and science continues its inexorable advance into understanding the mental and the emotional world, I hope that humanity will slowly be able to take off its mental straitjacket and move forward ever more freely to finding its own meaning, with fuller and more efficient use of its brain and Mind. This will be made even more effective by teaching people better understanding and control of the 'baser' emotional aspects of their lives, as outlined briefly above.

But we still need to discover how people might be able to give meaning to their lives. In terms of what I have written so far in this book, my answer to this very hard problem is as follows.

Through a careful education of the young in both emotional and intellectual abilities they will grow up knowing both what they like to do and why. The 'what' gives intellectual bite, the 'why' gives some sense of security, so that they do not feel out of control. Moreover, their likes and dislikes will have been developed in a caring manner so that they do not enjoy outrageously antisocial acts. At all times they would see themselves, and be seen by others, as in control of their own destinies, in pursuits which they have chosen because these are found to be satisfying to the emotional needs of the person. They will thereby become more self-aware and enhance their lives accordingly. They may still believe in a religion, being unable to accept the ultimate lack of meaning to life shown by science. That will always be their prerogative. I claim that is all life can ever hope to offer any human – to satisfy the internal goals and drives, both inherited and environmentally developed, in a manner which simultaneously allows others in society to achieve such satisfaction for themselves. The fear of suicide of the human race as it comes to realize the apparent meaninglessness of life (as I mentioned earlier) is, I would claim, a misapprehension. Suicides occur when a person can see no solution to the problems besetting them. Through emotional education and guidance, with better self-knowledge thereby generated, such 'end-of-the-road' responses should be reduced. Man and woman must understand themselves and each other better, at the deepest intellectual and emotional level, to achieve their ultimate humanity. Each must heighten their self-awareness and the awareness of others. As I see it, humankind has a better future through science and its applications. The meaning of life will then become clear – life is to be lived through to the full, both inside oneself and with others.

GLOSSARY

Amygdala A sub-cortical brain nucleus, adjacent to the hippocampus, which appears to be crucial for emotional memory.

Cerebellum The 'little brain' at the back of the brain stem involved in fine control of movement.

Cerebral cortex The surface of the brain, divided into two halves, composed of almost the same architecture of six layers of cells throughout. It may be sub-divided into primary regions where inputs arrive (visual, auditory, sensory, olfactory), or outputs leave (motor) or associative, where those inputs are analysed and related together or motor programmes are stored.

Frontal lobes Cerebral cortical region involved in planning and in assessing features of time (delays etc.), as well as in motor control.

Gluon The particle 'glue' (particle or quantum of energy) analogous to the photon, whose exchange between protons, neutrons or their constituent quarks can lead to the very strong binding known as the nuclear force, and to the binding of the quarks together to make the neutron and proton and their heavier companions.

Graviton The hypothetical particles acting as the source of the gravitational force. So far it has only been observed very indirectly.

Hippocampus Shaped like a sea horse, this organ is at the edge of the cortex and is essential for laying down long-term memory.

Hypothalamus A set of sub-cortical nuclei involved crucially in the perception of pleasure or pain, and more generally in the transfer of body-drive levels (thirst, hunger, etc.) into associated neural motivational states.

Limbic circuit A ring of groups of nuclei apparently critically involved in assessing the emotional significance of bodily (from hypothalamus) or external (from cortex) inputs.

Meson An epithet for a class of particles, originally thought to be the glue of the nucleus, but now recognized either as distinct fundamental constituents of matter (the mu-meson or the tau-meson, or W and Z mesons) or composed of combinations of pairs of quarks.

Neutron Almost identical in mass and other properties to the proton except for being electrically neutral; it is the other fundamental constituent of all nuclei besides the proton.

Nucleus Reticularis Thalami A sheet of inhibitory cells draped over each thalamus and acting as a gate to control relayed information between thalamus and cerebral cortex.

Occipital lobe Is at the back of the cortex, and involved crucially with analysis of visual input. It consists of numerous areas.

Parietal lobe Occupies a dorsal posterior position on the cortical surface, and is interposed between sensory and visual inputs.

Photon The packet or quantum of energy in light or other electromagnetic radiation, emitted or absorbed by charged particles; the exchange of photons between charged particles leads to the electromagnetic force first described in unified fashion by J. C. Maxwell in 1864.

Proton The nucleus of the hydrogen atom, possessing one positive unit of electric charge. It is a fundamental constituent of all matter.

Quark One of the triplet of fundamental constituents of protons and neutrons. There are several different sorts of triplet (with epithets up, down, strange, charm, top, bottom (or truth and beauty)), there being three pairs of these triplets belonging to each of the three known families of fundamental particles.

Spin states An intrinsic property possessed by particles making them similar to a spinning top. Matter (electrons, neutrinos, protons, neutrons) can all spin either upwards or downwards, while glue (photons, gluons, W and Z particles or gravitons) can spin in three directions (up, down or sideways). However only two of these states is seen if the glue particle has zero rest mass.

Temporal lobe Cerebral cortical region between visual input and long-term memory storage in the hippocampus; is in medial and lateral posterior positions on the cortical surface.

Thalamus Brain organ shaped like an egg, situated at the top of brain stem and acting as a relay centre for all inputs (except smell) to and from the cerebral cortex; each hemisphere has its own thalamus.

W and Z mesons Particles which act as the glue of the radioactive force, which is produced by their exchange. They are nearly one hundred times as massive as the proton, and have electric charge equal either to that of the proton or to the opposite of it (for the W), or zero (for the Z).

BIBLIOGRAPHY

I LIST HERE those books which I have found to be of especial importance in the writing of this book. There are a vast number of others on the topics covered, but I do not feel it necessary to list them since they did not seem to clarify my thoughts further on the questions I have tried to answer.

Adams, D., *The Restaurant at the End of the Universe*, London, Pan Books, 1980

Appleyard, B., *Understanding the Present: Science and the Soul of Modern Man*, London, Pan Books, 1992

Aristotle, *Metaphysics*, London, J. M. Dent, 1956

Baars, B. J., *A Cognitive Theory of Consciousness*, Cambridge, Cambridge University Press, 1987

— and W. P. Banks (eds), *Consciousness and Cognition*, London, Academic Press, 1992

Ballentine, L. E., 'The Statistical Interpretation of Quantum Mechanics', *Reviews of Modern Physics* 42, 358–81, 1970

Bergson, H., *The Two Sources of Morality and Religion*, Part II, Trs. R. A. Audre and C. Brereton, London, Greenwood Press, 1935

Cotterill, R., *No Ghost in the Machine*, London, Heinemann, 1989

Davies, P. C. W., *The Mind of God*, New York, Simon & Schuster, 1992

Dawkins, R., *The Selfish Gene*, 2nd edn, Oxford, Oxford University Press, 1989

Dennett, D. C., *Consciousness Explained*, New York, Little, Brown, 1991

Edelman, G., *Bright Air, Brilliant Fire: On the Matter of the Mind*, London, Allen Lane, 1992

Efstathiou, G., 'Mud-wrestling with COBE', Physics World 5, 27–30, 1992

Farley, J., *The Spontaneous Generation Controversy from Descartes to Oparin*, Baltimore, Johns Hopkins University Press, 1977

Forrest, S. and G. Mayer-Kress, 'Genetic Algorithms, Nonlinear Dynamical Systems and Models of International Security', in *Handbook of Genetic Algorithms*, L. Davis (ed.), London, Pitman, 1991

Gough, C., 'Challenges of high-Tc', *Physics World* 4, 26–30, 1991

Green, M. B., J. H. Schwartz and E. Witten, *Superstring Theory*, Vols I and II, Cambridge, Cambridge University Press, 1989

Hey, A. and P. Walters, *The Quantum Universe*, Cambridge, Cambridge University Press, 1987

Holland, J. H., 'Genetic Algorithms', *Scientific American* 267, 44–51, 1992

Honderich, T., *Determinism*, Oxford, Clarendon Press, 1981

Isacson, R. L., *The Limbic System*, New York, Plenum Press, 1976

Jerison, H. J., 'On the Evolution of Mind', Ch. 1 in *Brain and Mind*, D. Oakley (ed.), London, Methuen, 1985

— 'Palaeoneurology and the Evolution of Mind', *Scientific American* 19, 90–101

Krauss, L. M., *The Fifth Essence: the Search for Dark Matter in the Universe*, New York, Basic Books, 1990

Krick, F. and C. Koch, 'Towards a neuro-biological theory of consciousness', *Seminars in the Neurosciences* 2, 263–75, 1990

Kulli, J. and C. Koch, 'Does anaesthesia cause loss of consciousness?', *Trends in Neurosciences* 14, 6–10, 1991

Leslie, J., 'Efforts to Explain All Existence', *Mind* 87, 181–97, 1978

Linde, A. H., 'Particle Physics and Inflationary Cosmology', *Scientific American*, 36–43, March 1992

Lucas, J., 'Minds, Machines and Godel', *Philosophy* 36, 120–4, 1961

Madler, C. and E. Poppel, *Naturwissenschaften* 74, 42–3, 1987

Mayes, A. R., *Human Organic Memory Disorders*, Cambridge, Cambridge University Press, 1988

Medawar, P. B., *Limits to Science*, Oxford, Oxford University Press, 1984

Misner, C., K. Thorne and J. Wheeler, *Gravitation*, Oxford, Freeman, 1973

Peat, F. D., *Superstrings*, London, Macdonald, 1988

Peebles, P. J., D. N. Schramm, E. L. Turner and R. G. Kron, 'The case for the relativistic hot Big Bang cosmology', Nature 352, 769–76, 1991

Penrose, R., *The Emperor's New Mind*, Oxford, Oxford University Press, 1989
Polkinghorne, J. C., *The Way the World Is*, London, Triangle Press, 1983
Ponnamperuna, C., *Cosmochemistry and the Origin of Life*, D. Reidel, 1983
Popper, K. R. and J. C. Eccles, *The Self and Its Brain*, Springer, 1977
Rescher, N., *The Limits of Science*, Berkeley, University of California Press, 1985
Ribary, U., A. A. Ioannides, K. D. Singh, R. Hasson and R. Ilinas, 'Magnetic Field Tomography of Coherent Thalgamo-Cortical 40-Hz Oscillation in Humans', *Proceedings National Academy of Science*, 1987
Sacks, O., *The Man Who Mistook His Wife for a Hat*, London, Duckworth, 1985
Schachter, D., 'Toward a Cognitive Neuropsychology of Awareness: Implicit Knowledge and Anosognosia', *Journal of Clinical and Experimental Neuropsychology* 12, 155–78, 1990
She, Z-S and E. Jackson, 'Turbulence Called to Order', *Physics World* 4, 19–20, 1991
Schiebel, A. B., 'Anatomical and Physiological Substrates of Arousal: A View from the Bridge', 55–66 in *The Reticular Formation Revisited*, J. A. Hobson and M. A. Brazier (eds), New York, Raven Press, 1980
Schlegel, R., *Completeness in Science*, New York, Appleton-Century-Croft, 1967
Smith, A., *The Mind*, London, Hodder & Stoughton, 1984
de Solla Price, D. J., *Little Science, Big Science*, New York, Columbia University Press, 1965
Squire, L. R., *Memory and Brain*, Oxford, Oxford University Press, 1987
Stewart, I., *The Problems of Mathematics*, Ch. 15, Oxford, Oxford University Press, 1987
Taylor, J. G., 'Can Neural Networks Ever Be Made to Think?', *Neural Network World* I, 4–11, 1991
—— 'From single neuron to cognition', in *Applied Neural Networks* II, I. Aleksander and J. G. Taylor (eds), New York, Elsevier, 1992
—— 'Temporal Processing in Brain Activity', in *Complex Neuro Dynamics*, J. G. Taylor, E. Caianiello, J. Clark and R. Cotterill (eds), London, Springer, 1992

Taylor J. G., P. C. Bressloff and A. Restuccia, *Finite Superstrings*, World Scientific, 1992

Toates, F., *Motivational States*, Cambridge, Cambridge University Press, 1988

Trefil, J. S., *From Atoms to Quarks*, London, Athlone Press, 1980

Walsh, D., R. F. Carswell and R. J. Weymann, *Nature* 279, 381, 1979

Weinberg, S., *The First Three Minutes*, London, André Deutsch, 1977

Wanquier, A., and E. T. Rolls (eds), *Brain Stimulation Reward*, Amsterdam, North-Holland, 1976

Young, P. J., J. E. Gunn, J. A. Kristian, J. B. Oke and J. A. Westphal, *Astrophysics Journal* 241, 507, 1980

INDEX